Gödel's Mistake

Gödel's Mistake

The Role of Meaning in Mathematics

Ashish Dalela

SHABDA
PRESS

Gödel's Mistake—The Role of Meaning in Mathematics
by Ashish Dalela
www.shabda.co

Published by Shabda Press
www.press.shabda.co
ISBN 978-81930523-1-0
Second Edition
v1.5(06/2021)

Pure mathematics is, in its way, the poetry of logical ideas.
                                            —Albert Einstein

Contents

List of Figures

Preface

Gödel's Incompleteness Theorem is amongst the best-known results in 20th century mathematics and perhaps no other theorem has had such a great impact on such a wide variety of academic disciplines. The theorem proves that any mathematical theory about numbers will be either incomplete or inconsistent. Incompleteness implies that some true statements cannot be proved. Through the course of this book I will demonstrate that this incompleteness arises because numbers—pretty much like words in ordinary language—can denote both universals or concepts and particulars or individuals. There is no way to distinguish between these modes of using words or numbers, because mathematics lacks in grammatical categories like common nouns and proper nouns. Therefore, we must assume that mathematics either deals with concepts or with individuals, and we cannot mix them in the same sentence. Those propositions that involve both modalities can produce contradictions. But if you keep these modalities separate, then you cannot prove or disprove the propositions that mix them. To avoid some contradictory situations that arise from mixing them, we must take the precaution of using only one mode in all the cases (even when there is no contradiction). This precautionary measure leads to *incompleteness*—namely that you cannot prove or disprove some propositions. But if you tried to overcome this issue, then this would lead to contradictions.

Based on this diagnosis of the problem, the solution is also evident—mathematics should be treated as a language, which it already is, but not just a special language as it is treated right now. It should rather be treated on par with ordinary languages, meaning figures of speech that exist in ordinary language must exist in mathematics too. This is not a light baggage to bring in, because adding figures of speech introduces grammar, the use of the same word to represent different meanings (universals or particulars) introduces contextuality, and all the problems that exist in understanding ordinary language now enter mathematics.

We can also put the problem in reverse—instead of trying to make mathematics like ordinary language, we can try to mathematize ordinary language. At the least, a deeper understanding of ordinary language, and its differences from current mathematics, can pave the way to the understanding of where the real future of mathematics lies. In so far as the understanding of ordinary languages has daunted linguists for centuries, we can say that the problem is hard. Conversely, in so far as current mathematics is successful in whatever it is doing right now, it is immensely incomplete.

Since practically all areas of modern science use numbers, Gödel's theorem implies that all these fields of science are incomplete and that they will contain truths that cannot be proved. I will not delve into this topic directly in this book, but in my other work I have demonstrated that the pattern of inconsistency vs. incompleteness repeats in every area of science. In atomic theory, for example, it appears as Bell's Theorem where the theory must necessarily remain probabilistic—i.e. make incomplete predictions—because trying to complete it by adding some hidden variables will make it contradictory. Similarly, in computing theory—and this is something that I will discuss later in this book—you cannot decide if a program will halt, and therefore whether it is useful or malicious in any automated way. If, however, you suppose that such an automated procedure could exist, the assumption results in a contradiction. Accordingly, computing theory must be incomplete. The problem of incompleteness takes many forms in different areas of science, but the attempt to overcome it results in the shared consequence of a contradiction. Therefore, I contend that these are all problems associated with the difference between how we describe nature using mathematics and how we do so in ordinary language.

Quite specifically, in science and mathematics we have chosen to describe the world as *objects*. We do use concepts in forming theories, but every concept is eventually reduced to an object. For example, the property of 'mass' is a concept, but we always reduce it to a measuring instrument such as a kilogram or pound, thereby avoiding the need to deal with a concept, and only deal with a *value*—the outcome of a measurement against the chosen standard. We continue to think that material objects 'have' mass, but we don't have to define what it is; we just compare it to a chosen standard object. Therefore, science never really has to deal with mass. It only deals with a chosen standard which becomes a *representation* of the concept. That chosen standard is an object, rather than a concept. Which means that you can always substitute every concept for a value against an individual object. By that clever trick, mathematics avoids all concepts.

The trick, however, fails when we try to define numbers, because a number is indeed a concept—which can be found in many collections of objects. You should not ideally tie the number five to a specific collection of five objects, as you could do with a property like mass (by tying it to a standard object like a kilogram). But mathematicians tried that anyway. They argued that we can suppose that five is indeed like mass, and just as mass can be reduced to a kilogram similarly we can reduce five to a specific collection of five things. The problem, however, is that to do so you must know that the collection indeed has five things, not six or four. To ensure that the collection has those five things, you must count—one, two, three, four, and five. That counting requires the concept 'five' to exist before you can decide that there is a collection of five things. It follows that 'five' must exist as a concept before you can form a collection of five things, so you need the concept to define the concept.

This issue reintroduces the original problem we were trying to avoid, namely, to use two modes—concepts and individuals. There have been many attempts to avoid this problem, which constitute an interesting history about the foundations of mathematics that we will discuss in this book. But, to give away the mystery which lies at the end of this story, all of them have failed. If numbers are the most fundamental objects mathematics deals with, then we don't have a foundation of mathematics because we need to define numbers as concepts (even to define a numbered collection of things), and the introduction of concepts introduces two modes in language, and to deal with two modes we need figures of speech, which then needs a grammar, and a whole slew of problems immediately follow.

Therefore, although the problem can be diagnosed easily, and I will do that in this book, the solution to that problem requires a radical revision not just to mathematics, but to everything that uses mathematics—the use of multiple *modes* of language, followed by grammar and contextuality. Pending that radical revolution, there can be no foundation of mathematics, and every area of science is incomplete. But if you try to fix that incompleteness you will end up in a contradiction. Gödel's incompleteness therefore has immense ramifications for every area of modern rational inquiry.

In fact, as we will see during the book, the solution to the problem also involves revisions to logic because when we get down to defining concepts, we find that they always come in opposites. For instance, if 5 is a concept, then -5 is also a concept. Indeed, they can only be defined mutually or not at all. The universe of these concepts can't be consistent; that is, in the universe where 5 exists, -5 will also exist. So *collectively* the conceptual universe is inconsistent. We already had problems defining concepts, but we did not have a problem defining the collection of 5 things assuming the concepts were defined. A new problem now appears in defining negative numbers—we can't even make a collection of -5 objects. To do that, we would have to be measure the *absence* of things rather than their presence. This means that you can no longer reduce -5 to an object. You must explicitly introduce concepts or *types*—some that are detected by presence and others that are detected by an absence. This naturally doubles the number of concepts required.

Problems also appear in defining rational and irrational numbers, but this time we find many cases in which we can draw a physical instance—e.g. a circle representing π or a hypotenuse of an equal sided right-angle triangle representing $\sqrt{2}$—so there is a physical instantiation of these numbers, but how do you represent them conceptually given that they often have infinite digits?

Recall that the problem of Gödel's incompleteness arises if we use both universal and individual modes in the same sentence. So, whether the problem pertains to the conceptual definition of numbers or to the physical instantiation of these numbers, the incompleteness stands. Therefore, to make mathematics—and the rest of science complete—we need to address both types of definitions. Unless we do that, mathematics doesn't have a foundation because we aren't able to define what number is, regardless of how much sophisticated arithmetic, algebra, or geometry we make out of it. Exploring this incompleteness and pointing toward a possible solution to this problem is the prime aim of this book.

There are two ways to interpret incompleteness. We can treat incompleteness as an indication of uncertainty in the world and that we can't know the truth of some statements because they don't necessarily exist in a true or false state. Alternately, we can regard incompleteness as the uncertainty in our knowledge, which could be seen as the human inability to know all things, potentially a limit of our mental capacity. In either case, it means that mathematics cannot describe nature's entire splendor because the language in which we describe problems and their solutions is partially incapacitated.

Of course, not everyone takes Gödel's limitations so seriously. You might for instance claim that limitations to mathematics do not entail a shortfall in our knowledge because the methods by which we acquire knowledge go beyond mathematics. In particular, if logic is inadequate to obtain knowledge, we may use sensation, or intuition. This may be a good approach as a theory of knowledge, but it doesn't really solve the problem because the knowledge we acquire through sensation or intuition must be expressed in language as much as logical knowledge. The problem of mathematics is thus not a problem of logic, but of the language in which we describe reality. While sensations or intuitions may solve the problem of knowledge acquisition, they won't solve the problem of consistently *describing* this knowledge in a way that can be communicated to others.

This book therefore sidesteps various approaches to obviate the serious implications of Gödel's incompleteness, because these approaches focus on knowledge acquisition rather than on the problem of knowledge *expression*. Instead, I will focus on identifying some problems with Gödel's proof of incompleteness. These problems are conceptual and not logical, in case you are wondering how such a well-known result could have issues within its proof.

Gödel's proof uses three modalities about numbers—a thing, a name and a concept—and the proof rests on being able to use them interchangeably. Once this equivalence between the three uses of a number is dropped, Gödel's proof of incompleteness does not exist. This means that if we were able to distinguish between name, concept and thing interpretations of a number, mathematics could be complete. It also means that numbers themselves have many semantic interpretations, which most mathematicians find surprising because they treat numbers as quantities. Gödel's theorem shows that there are three meanings of numbers within mathematics, although mathematics does not have ways in which to distinguish between them. The confusion between the three uses of numbers leads to logical paradoxes. If mathematics could distinguish between these three meanings, Gödel's paradox would not exist.

One of the best-known undecidable problems today is Turing's Halting Problem which states that mechanical procedures cannot determine if a program will halt. Like Gödel's theorem, Turing's proof also relies on different uses of numbers although, unlike Gödel's theorem, which treats a number as a thing, concept and name, Turing's proof uses numbers as data and programs. If there was a way in computing theory to distinguish between data and programs, then Turing's proof would not exist. I will show that this distinction requires the notion that a program solves some *problems* and the problem is distinct from the program. The program (as an algorithm), further, has to be distinguished from the physical bits.

Most mathematicians don't recognize the important role meaning plays in mathematics and computing. This book discusses how meaning enters mathematics through different interpretations of numbers. The need to distinguish between these interpretations represents a general class of problems not limited to mathematics. These problems have been encountered in linguistics as the conflict between universals and particulars. They appear as the difficulty in distinguishing body and mind in psychology. The problem also has important implications in physics when matter is used for semantic computations. I take on these latter implications in another book, *Quantum Meaning: A Semantic Interpretation of Quantum Theory*.

The pervasiveness of the problem implies that a solution to the problem of semantics in mathematics will have consequences for computation, physics, psychology and linguistics as well. This pervasiveness also means that there is something fundamental missing in our understanding of languages and how these languages represent meanings and this missing ingredient must be understood and incorporated in our viewpoint about language before problems of incompleteness can be appropriately solved within mathematics.

The missing ingredient in mathematics is the contrast that it bears with respect to everyday language. Multiple ways of counting are inherent in everyday language. An ordinary word like 'table' can denote the name by which an object is called, a meaning in our mind, or things in the world, and we seem to naturally understand how to distinguish between names for identifying, mental meanings and the physical things. Ordinary language also allows the same word to be used as a concept and an action, or a noun and a verb, and we have learnt how to distinguish nouns from verbs (a word such as 'address' or 'color' can be a noun and a verb, depending on the context).

While ordinary language uses words in many ways, different methods of giving meanings to numbers don't exist in mathematics. Gödel's incompleteness, Turing's Halting Problem and other logical and mathematical paradoxes that I will survey in this book result from this difference between mathematics and everyday language. If we cannot distinguish between a thing, a name, a concept and a program, then logical contradictions can be constructed by interpreting one meaning of a number as another. To fix incompleteness, we must fix the shortfall in mathematics vis-à-vis ordinary language. Fixing the shortfall entails the ability to distinguish between various interpretations of numbers, but that also brings in grammar. Once the ability to distinguish between different meanings of number has been incorporated within mathematics, then the paradoxes in mathematics will not exist. This helps us understand why the paradoxes of mathematics don't arise in the everyday world because the everyday world incorporates the distinctions in ordinary language. Sensations or intuitions are therefore viable means to decide on the truth of a claim because in sensation and intuition we use everyday language that incorporates distinctions which don't currently exist in mathematics. The same ability can also exist logically in mathematics, if mathematics incorporates the distinctions currently seen in ordinary language. Such a mathematical theory of numbers can also be a rational explanation of the working of sensation and intuition.

The reader will find here new connections between computation, mathematics, philosophy, logic and cognitive science. These point towards the relation between mathematics and a semantic theory of computation. The book draws upon insights from the nature of ordinary language to sketch changes to mathematics and what they mean for theories on computing and the mind.

Ordinary language has a lot to tell us about the human mind, the nature of reality and computation. This might be the most surprising conclusion of all—that as a descriptive language, mathematics lags behind ordinary language. But this might not, after all, be that surprising if we recognize that ordinary language is more powerful than mathematics, and that we don't yet fully understand everything that gives ordinary language its powers. By understanding ordinary language and its differences vis-à-vis mathematics, and then inducting those distinctions of ordinary language into mathematics, it is possible to bridge the chasm between the logical world of mathematics and the world described by ordinary language.

1

Mechanizing Thought

*Either mathematics is too big for the human mind or the
human mind is more than a machine.*

—*Kurt Gödel*

Introduction

After the numerous successes of industrialization in the 18th and 19th
centuries, there came attempts in the 20th century to mechanize the
mind. After all, if a machine can input cotton and output yarn, why
can't it input symbols and output theories and theorems? Leading
the bandwagon to automate the mind were mathematicians. Proving theorems has never been easy, so why shouldn't we delegate this
job to machines? Given any kind of problem, the machine would be
tasked to search[1] for long hours until it stumbles upon a proof. Since
machines can work much faster and longer hours than humans, they
could one day perhaps prove theorems much quicker than humans. A
machine churning out theorems, with only electric power supply as
input would be such a wonder! But, while the idea is indeed fascinat-
ing, it is also perplexing that mathematicians should pursue it. Unlike
the union wars of the 19th century where laborers fought against the
induction of machines that rendered them jobless, here was a set of
very intelligent and successful people hell-bent on automating their
profession. In a classic twist of desire, mathematicians wanted to
commoditize the very task of proving theorems that had made them
socially significant and attractive.

Of course, a significant amount of groundwork had already been
done to make mathematicians believe that this was imminently

1

possible. Logic and the methods of reasoning had been systematized during the time of Aristotle. Euclid had formalized geometry through a collection of five axioms. Charles Babbage had drawn up mechanical designs for carrying out calculations in the late 19th century. George Boole, through his pompously titled book *Laws of Thought*, had made significant progress in converting mathematical operations into logic, most of which underlies the technology used in the Arithmetic-Logical Unit (ALU) of present-day computers. Gottlob Frege had taken strides in bringing ordinary language statements closer to logic. There was hence a widespread belief in the rationalist academia of early 20th century that mathematics and ordinary language are different sub-branches of logic and can be derived from it. Under the circumstances, it seemed that a machine that would think exactly like human beings was just a matter of time.

Hilbert's Second Problem

Every time we set out to solve a problem, it is helpful to know if the problem is, in principle, solvable. Otherwise, we may be wasting our time trying to solve what is, in principle, unsolvable. In the context of mathematics, mechanizing theorem proving needed the proof that everything that we set out to prove can be provable (or disprovable). For, otherwise, a machine would be trying to prove something that can never be proved, a veritable waste of time. The idea that everything in mathematics can be proved or disproved is a meta-mathematical question and if this idea were true, then mathematics would be both *consistent* and *complete*. Practically every mathematician in the early 20th century believed that mathematics is consistent and complete, although it had never been formally proved. How do you show that every possible statement in mathematics is provable without actually going through the proofs of all of those statements individually? In his famous lecture at the International Congress of Mathematicians in Paris in 1900, Hilbert stated 23 problems, out of which proving the consistency and completeness of mathematics was called the second problem. Hilbert wanted to formalize it as an explicitly stated problem, perhaps to divert attention, funds and research effort towards it. He noted[2]:

"Upon closer consideration the question arises: Whether, in any way, certain statements of single axioms depend upon one another, and whether the axioms may not therefore contain certain parts in common, which must be isolated if one wishes to arrive at a system of axioms that shall be altogether independent of one another. But above all I wish to designate the following as the most important among the numerous questions which can be asked with regard to the axioms: To prove that they are not contradictory, that is, that a definite number of logical steps based upon them can never lead to contradictory results."

Hilbert's second problem has two parts. The first is about ensuring that we have unique and non-overlapping axioms using which any statement can be proved or disproved. The second is that this set of axioms must be consistent. If there is a set of axioms using which any mathematical statement can be proved or disproved, then logic is enough to decide the truth of all statements, assuming the axioms are true. The only imaginable hurdle to such a method of validating the truth of any claim would arise if the use of logic results in a contradiction between the axioms themselves. To avoid this unpleasant alternative, it was important to discover the axioms such that they will never contradict each other and then prove that they are consistent and complete. If an axiom set is shown to be consistent and complete, anything logically derived from those axioms should also be consistent and complete and mathematics as a whole will thus be consistent and complete. At least, that's how it seemed.

Of course, we assume that logic itself is consistent and complete and that the application of logic to prove or disprove statements will not create contradictions. Although we take it for granted, this assumption isn't trivial and needs to be independently validated. There are then two important steps to demonstrate the completeness and consistency of mathematics. First, we must prove that logic itself is consistent and the use of logic will not create inconsistencies. Second, we must show that there exists a complete set of axioms using which every statement can be proved or disproved. Given a complete set of axioms and rules of logic to mutate axioms into statements, we can prove the consistency and completeness of mathematics; logic and axioms will be sufficient for mathematics.

Gödel proved that logic is consistent and complete, and this is called Gödel's Completeness Theorem. He then showed that it is impossible to have a consistent and complete set of axioms for arithmetic and this is called Gödel's Incompleteness Theorem. Gödel showed that any axiom set for arithmetic could be either consistent or complete, but not both. The result of the Incompleteness Theorem is that any theory that deals with numbers will either prove some false statements or will not prove some true statements.

With the Incompleteness Theorem, an endearing dream of the rationalists was shattered. To understand its impact, it is important to realize the special position that mathematics had enjoyed amongst sciences and why it was very important to reinforce that position. The need to prove the consistency and completeness of mathematics goes back to a quarrel between rationalists and empiricists over which constitutes a superior method of knowledge. Rationalists (mathematicians being the prominent ones) claim that reason is superior, if we can formulate all necessary axioms. Experience, they claim, is fraught with hallucinations, subjectivity and perceptual errors and these errors must be removed before experience can constitute sound grounds for knowledge. But whatever method you use to correct problems with perception (such as repeating an experiment over and over again to validate its truth), the method also suffers from the same limitations as the original perception itself. Reason, on the other hand, does not suffer from perceptual limitations and is therefore the correct approach to knowledge.

The empiricists too had a challenge for mathematicians. They argued that the axioms from which mathematicians prove theorems are generalizations of everyday experience. For instance, if we did not see multiple objects in the world, we would have no need to count them, and the idea of number would not need to exist. Similarly, geometry is the generalization of the experience of space. Notions such as irrational and rational numbers are further derived from experience. Therefore, if we did not have experiences then axiom generation would be impossible. Some philosophers have gone even further to argue that even logic is derived from experience. For example, the idea of mutual-exclusion in logic is based on the separability of things, or the ability to identify them individually. Similarly, the principles of identity and non-contradiction are based on the commonsense notion of how things are separate. Logic and mathematics have therefore been derived from experience.

The problem posed by empiricism, however, had become tenuous by the early 20th century because physicists were relying on mathematical theories that were developed without reference to any reality. Classic examples of these theories include the Hilbert Space formulation used in quantum theory and Riemannian geometry used in general relativity. The idea therefore that mathematics is derived by abstracting and generalizing observations seemed to be inverted. It was truer to suppose that mathematics is independent of reality, although mathematical theories are useful in modeling physical phenomena. This idea goes back to Plato in ancient Greek times who claimed that there is a pure world of forms above us, of which the present world is but a poor imitation. Most mathematicians have been Platonists. Even Bertrand Russell who was a Logicist and not a Platonist claimed that mathematics is true in every *possible* world while physical theories are only true in our present world.

Thus, the empiricist argument that mathematical axioms are derived from the observation of reality did not stick. Mathematicians claimed that knowledge of reality was the job of physics, not mathematics. The job of mathematicians was to find theorems given some axioms. Which axioms are relevant to the study of the real world was irrelevant to mathematicians. In particular, it was possible to believe that mathematics can postulate axioms even when they violate physical intuitions and so mathematics did not have to depend on experience. Classic examples of such entities are imaginary numbers which can never be physically computed, so there was no empirical basis for them. To extend this idea to all of mathematics, it was important to demonstrate the autonomy of mathematics relative to other fields of science. The ability to construct arbitrary axioms and prove that they are consistent and complete was central to the genesis of new mathematical fields.

Thus, while no one had actively disputed the completeness and consistency of mathematics, mathematicians felt it important to prove it explicitly. It was important to continue doing mathematics in the mind and to show that it is a system of knowledge that does not depend on reality. Hilbert specifically looked upon mathematics as the ability to create well-formed sentences, whether these sentences describe reality or not. The essential criterion for something to be true in mathematics is logical consistency with axioms, and not a correspondence between ideas and reality. Mathematics could be independent of worldly facts, and its practice need not involve anything outside the human mind. This would establish mathematics as the queen of all sciences that provides the basic tools of thinking. We might now say that mathematical ideas are Platonic.

The philosophical irony of Platonism is that the mind by which we do mathematics is also in some sense material. The mind is certainly not other-worldly and the creation of symbols, theories and theorems in the mind depends on the laws of nature. Mathematicians can pursue the search of possibility because that possibility is permitted by reality. For instance, if the laws of nature forbade the construction of some sentences such that they could not arise in the mind, then they could not be discovered by mathematicians. So, by pushing the problem from the external world into the mind, the problem doesn't disappear. It just gets postponed a bit more. Theorems are also *things* that *can* exist in matter, and all possible theorems are real in the mind because they can be represented as physical states in the mathematician's mind-brain. As a material object, the mathematician's brain is constrained by the laws of nature, and those laws must explain the mathematician's ability to formulate theorems. There is hence a deeper sense in which mathematics is not independent of matter that mathematicians shudder to acknowledge. This dependence between ideas and matter appears as a limitation of Platonism, as we shall see later.

As it turns out, Hilbert's program to prove the consistency and completeness of mathematics was stopped in its tracks by a proof that contradicted the possibility of fulfilling his dream. The person instrumental in destroying Hilbert's dream was Kurt Gödel who showed that mathematics can be either consistent or complete, but not both. Gödel's proof assumes that all meaningful statements are either true or false, and then shows that this assumption is inconsistent. It follows that all meaningful statements aren't necessarily true or false, or that we could not know their truth status. Mathematically, a meaningful statement is one that is well-formed according to the rules of correct grammatical construction. The statement is further true if it can be *derived* from the axioms or premises. It is tempting to believe that in a purely formal language like mathematics, every statement that is grammatically correct can also be proven or falsified logically from the premises.

But this is a key point of inflexion in Gödel's theorem because truth is not just logical but also empirical. To be meaningful, the statement must be grammatically correct, but its truth may not be logical; the truth could also be empirical. Given any random statement, it is impossible to know upfront if its truth should be derived logically or empirically. We cannot logically prove statements whose truth is empirical. Mathematicians assume that all statements in a logical discipline must have a logical truth, but Gödel formed a type of statement that refers to other statements. The truth of such a statement is expected to be known empirically. But, through a trick, Gödel converted the referential statement to a self-referential statement; that is, it was not necessary to look at another statement to know its truth. If the self-referential statement involves a contradiction, then mathematical theory also has contradictions. Such statements cannot be proved or disproved within the theory, and the mathematical theory is now deemed incomplete.

Gödel was the first to demonstrate incompleteness in arithmetic but several people before him had used similar ideas in philosophy and logic. Before I examine Gödel's proof, it will help to look at the history of ideas that preceded Gödel's theorems. Ideas that Gödel employed in his incompleteness proof did not begin with him. The problem of this theorem is so generic that it had existed before Gödel and has appeared after Gödel as well. The discussion of these problems will help illustrate the root issue and I will show that the root issue transcends mathematics and can be seen in philosophical and linguistic debates that existed before Gödel's theorem. Before discussing Gödel's proof and its issues, let us, therefore, turn to the wider philosophical climate that existed prior to these theorems.

Frege's Distinction

The problem of the relation between meaning and truth (that all meaningful statements may not be true) was known to philosophers before
Gödel's theorem although in a different sense. Gödel's proof brought
to the fore the *logical* meaning-truth distinction in mathematics while
philosophers before Gödel were concerned with the *empirical* meaning-truth distinction. The contrast between meaning and empirical
truth can be shown by a simple example.

Assume that we have a statement that must be proved empirically.
To confirm the statement, there must be some physical phenomena
that the statement describes or explains. If the statement has no real-
world application, then it cannot be empirically proven, and it remains
a mental construct, which is logically true, but with no practical relevance. We might liken it to a dream that is logically consistent but has
no counterpart in physical reality. The criteria that make dreams false
would make the statement false too. We might say that the statement
is logically true but it is empirically false, because there are no objective facts in the world that such a statement succinctly represents. For
instance, the term 'golden mountain' denotes a meaningful idea that
does not exist. And yet, we cannot claim that golden mountains will not
exist. Statements about such a concept (e.g., "The golden mountain is
very high") cannot be proved, since proof requires the 'golden mountain' which does not exist. And they can't be falsified either, unless we
can be sure that they don't exist. Statements about the 'golden mountain' however are perfectly meaningful and can be understood. And
given an empirical notion of truth, this is a straightforward contradiction of the idea that all meaningful statements must also be provable. If we form a theory that has 'golden' and 'mountain' as axioms
and given the rules of axiom combining that produce complex concepts from simpler ones, 'golden mountain' is a valid theorem derived
using the axioms 'golden' and 'mountain'. If we now try to validate this
theorem empirically, we will find that in a world where 'golden' and
'mountain' are true axioms, the theorem 'golden mountain' is not. And
this implies that even when valid axioms in a theory are being combined by logical rules for axiom combining, we can create a claim that
cannot be proven, even though the axioms are true.

But it is too early to concede defeat. Just because an idea does not have a worldly instantiation does not mean that there are no physical phenomena to which these theories or ideas will ever apply in the future. We just might not have found the right phenomena that need to use this mathematics or idea in a description. We might perhaps wait a little longer and search a little harder to find the phenomena that validate a theory or idea. Nevertheless, this is not an easy concession to grant. How long must we wait for the phenomena to be found? And until the time we have found the matching phenomenon should we regard a fact true or false? The best possible answer under the circumstances is just to say that the statement is neither true nor false, but simply not provable or falsifiable. We don't have any specific empirical facts to validate the theorem right now, although we might have had something in the past or we might discover something in the future that might prove or disprove it.

Expectedly, not everyone is going to be so charitable. John Stuart Mill for example said that the meaning of a statement is the facts it points to and if an idea does not correspond to facts, then it is meaningless. The view is obviously extreme because if indeed there was a 'golden mountain' that existed sometime in the past or could exist in the future, then that existence would instantly make the claim not just true but also meaningful, at that specific point in time! How could something be meaningless now and meaningful at another time? Are we to associate meaning with the passing of time? Mill's claim would also imply that some mathematical theorems that are meaningless today—because they currently don't have a real-world application—could be meaningful in the future, and that is an even more troublesome solution than the problem it tries to solve. The situation gets even more complicated when we start dealing with nebulous ideas such as good, liberty, justice or beauty because there is nothing directly corresponding to these ideas in the world, and it is hard to define their meaning accurately. By Mill's standard, these become meaningless, because they are not empirically true.

To avoid this extreme, Gottlob Frege instituted a distinction between *sense* and *reference*. Sense pertains to the intrinsic meaning of a statement while reference pertains to the objects the statement refers to, which in turn determine whether the statement is true or false. Frege suggested that a term that does not have a physical instantiation—like 'golden mountain'—is meaningful although false. Just because something does not exist in reality does not imply it is also meaningless because the meanings of statements are different from their truths. Frege's distinction is the claim that all meaningful statements are not necessarily true or false, which contradicts the view that Hilbert held, although with regard to empirical truth.

Frege's distinction between sense and reference has the germ of the possibility that Gödel eventually proved, namely that some meaningful statements may not be provable. According to Frege, the inability to prove the truth of a meaning isn't a problem because meanings are distinct from truths. The key distinction between Frege's idea and Gödel's theorem is that Frege was talking about empirical truth while Gödel dealt with logical truth. While Frege's conclusion is intuitively obvious, Gödel's theorem is not. Frege's conclusion is obvious because the everyday life is full of meaningful statements that are false. Every book on fiction is meaningful and yet most likely false. We understand that grammatically correct statements only give us meaning but not truth. The paradox in Gödel's theorem can therefore be solved if we can show that Gödel's proof also involves empirical truth rather than logical truth. This is something I will try to show later by demonstrating that Gödel's proof rests on the construction of *statements about statements.* Such statements are quite like the statements made about some reality; the reality in question now is not a physical object but a statement. Therefore, the truth of a statement about a statement involves an empirical notion of truth, which can't be proved using logic.

Logic and Existence

Frege's approach to language and meaning is disastrous for rationalists, because it introduces a role for experience in the knowledge of truth. Under Frege's approach, mathematics only deals in the meaningfulness of propositions, but not whether they are true. This was the main argument that empiricists had against rationalists when they claimed that truths about the world cannot be discovered through reason. Under this premise, non-Euclidean geometry was meaningful prior to the advent of relativity, but it became true only after the formulation of relativity. By instituting a distinction between sense and reference, Frege solidified the empiricist's argument against rationalism. Mathematics is not self-sufficient as it depends on observation. Couldn't there be a more amiable solution? To rescue mathematics from the compromising situation where it is seen as dealing with meaning and not truth, Russell came up with an idea. He suggested that it is possible to speak of the 'golden mountain' in a way that allows the truth to be verified by experience and yet does not surrender the burden of truth to experience. This way was obtainable if we introduce the notion of *existence* within logic.

Thus, a statement like "The golden mountain is very high" can be transformed into "There *exists* a thing, such that the thing is a mountain and it is golden, and it has the property of being high". What is existence? It is indeed empirical, because we are claiming that something exists in the world. And yet, by introducing existence within the statement, the meaning of the statement now points to reality. Logical inferences on this meaning can therefore appear to say things about reality, without having to consult facts in the world. What Russell said is a variation of what Mill had said earlier. Mill equated meaningfulness with facts and Russell says that the meaning of a statement is the *claims* it makes about the world.

For Mill, if something doesn't correspond to facts then it is meaningless. But, for Russell, if a statement does not correspond to facts, then it does not become meaningless, because the meaning is the claims it makes about the world. If the claims are false, then the statement is false, but not meaningless. Thus, if a statement is well-formed (conforms to the rules of grammar and other linguistic conventions), it will have an understandable claim about the world. The claim could however be true or false, but that is different from its meaningfulness. In Russell's scheme, statements are meaningful because they make some claims. If these claims are proven then the statement is true, and the meaning of statement is the facts it claims must exist for the statement to be true. Russell's scheme connects meanings with the world without reducing the meaning to facts. However this scheme is also saddled with an important problem.

Imagine that you are narrating a Harry Potter story. Are you stating claims about the world or merely conveying information? Given that Harry Potter is fiction you could not be stating claims about how things exist in the real world. To think that fiction becomes meaningful because it stakes claims is false because readers of fiction know that nothing in a work of fiction is true. But that doesn't deter readers of fiction from finding the fiction meaningful. Readers know that fiction is meaningful even when it is not a claim about the external real world. There can be meaningful things that are not true; but sometimes outright false things are also perfectly meaningful. It is thus wrong to convert all statements into claims about existence of facts to understand a statement's meaning. We could know their meanings even if things don't actually exist.

You might say that Russell's theory does not apply to fiction. But how do we distinguish a work of fiction from a theorem that does not have a physical application or at least hasn't found one yet? In effect, such a theorem is also a work of fiction—internally consistent and yet something that cannot be proved empirically. Russell's conversion of statements into claims therefore does not work as a theory of meaning. By suggesting that meanings are claims about truth, Russell allows the distinction between meaning and truth. However, by allowing Russell's view we would have to discard literature and poetry as meaningless, if they don't make truth claims.

There are further issues with Russell's theory, which arise when a statement does not make a definite claim at all times, although it does make different claims in different contexts. This problem with Russell's theory of descriptions was illustrated clearly by Saul Kripke[3]. To demonstrate, let us look at two descriptions of Aristotle. We might designate Aristotle as 'Plato's student' or as 'Alexander's teacher', both of which are empirically true. These two descriptions of Aristotle in turn lead to two statements: (a) Aristotle is Plato's student, and (b) Aristotle is Alexander's teacher. By the law of equals—since both statements pertain to Aristotle, they must be equal—the above two statements lead to another statement "Plato's student is Alexander's teacher". This statement is true in the specific case of Aristotle, but not true in general. If we assume that the statement means that "there *exists* a person who is Plato's student and Alexander's teacher" then the statement is true because Aristotle is indeed such a person. If however we treat the statement to mean that "every Plato's student is Alexander's teacher" then the statement is false because not all of Plato's students were Alexander's teachers, although Aristotle as one of the students of Plato was Alexander's teacher. Russell presumed that every statement will have a single interpretation or meaning and shall point to a single empirical fact. However, statements can also point to multiple facts and then claims about existence may not be true. The difference in meaning that comes from "one" and "all" becomes obvious when we reverse the subject and object in the statement. The reversed statements "Plato's student is Aristotle" and "Alexander's teacher is Aristotle" are both false as universal truths because Alexander had teachers besides Aristotle and Plato had students beside Aristotle. So, 'Aristotle' is 'Plato's student' but 'Plato's student' is not necessarily 'Aristotle'. Or, 'Aristotle' is 'Alexander's teacher' but 'Alexander's teacher' is not necessarily 'Aristotle'. Russell's use of existence works only for single claims but not with multiple claims in the same statement.

The only way we can get out of this problem is if we differentiate between the *meanings* of the word Aristotle from the *person* Aristotle. The meanings of Aristotle include being 'Plato's student' and 'Alexander's teacher' although the meanings need not always be equated with the person. More specifically, 'Plato's student' and 'Alexander's teacher' are mutually exclusive meanings of Aristotle. They are connected not because of a semantic relation between the meanings but because they apply to the same person. If we think of the person Aristotle as the physical instance that embodies the meanings 'Plato's student' and 'Alexander's teacher' then there is a clear distinction between the meaning and truth of the word 'Aristotle', as originally instituted by Frege. Keeping the distinction is however contrary to the spirit of Hilbert's program, which was the attempt to show that all meaningful things should be true or false.

No matter how we look at this, the distinction between meaning and truth is inevitable. It appears in the fact that every logically proven theorem may not find applications in the real world. This is demonstrated in the fact that mental ideas may not correspond to things in the world. And it is also shown because the meaning of a statement does not always tell us if that meaning is true or not.

In many ways, the distinction between meaning and truth was known outside mathematics, although mathematicians believed that this might not apply to them. Hilbert's program stated the overcoming of this problem as a goal. Inherent in this program is the belief that linguistic and philosophical distinctions between meaning and truth pertain to empirical truth while mathematics deals with logical truth. It did not occur to people that mathematics and ordinary language are both methods of framing sentences and that the problems of ordinary language could crop up in mathematics as well. There are two ways in which the problems of ordinary language could recur in mathematics: (a) by making statements about other statements, such that the truth of a statement is empirical and (b) if the relation between subject and object is reversed. Given that the truth of empirically referential statements cannot always be logically demonstrated, the theory is incomplete. And given that in the case of a reversal the claims may be false, the theory is inconsistent.

Hilbert's Formalism

The issues encountered in the philosophical analysis of ordinary language were not considered to be serious issues in mathematics for a couple of reasons. First, it was widely believed that the problems of language are issues with the metaphorical ways in which language is used, and words may not often have unambiguous definitions. Second, and more importantly, it was felt that the whole question of meaning and empirical truth was irrelevant to mathematics because the subject deals only with logical relations, not meaning.

Mathematicians—when faced with acrimonious questions about the role of meaning—often espouse a view that discards all questions about meaning in mathematics because they believe that mathematics is the exercise of manipulating symbols and understanding their interrelations. *Formalism*, as this view is called, is the idea that mathematics does not have mental meanings or physical references. A mathematical theory is not about ideas in the mind or objects in the world, although it does sometimes apply to both mental ideas and physical objects. Mathematics is rather purely a manipulation of symbols and some mathematical ideas are defined in terms of other mathematical notions, all the way up to primitive notions called *axioms*. The truth of the statement has got little to do with what this statement represents in the real world or what it means in the mind, if only the rules of symbol manipulation can construct the statement from the axioms alone. If the statement can be constructed, then it is true. And statements that cannot be constructed from axioms are false. If every statement is either true or false, then each statement must be built by applying the rules of logical deduction or the statement must explicitly contradict the statements that can be built from axioms. In either case, we would be able to prove or disprove them easily. Issues about reality and the mind therefore did not enter a mathematician's consideration.

With the eventual frustration of Hilbert's program, it is necessary to reexamine the Formalist notion that mathematics is a symbol manipulation problem. The reexamination however need not get into the question of mind and matter. In ordinary language, there are two ways to create a statement—by constructing a grammatically correct statement or by logically deriving it from other statements. The former sets the bounds on meaningfulness and the latter, on truth. Mathematicians had not distinguished the grammatical route to constructing statements from the logical route to deriving them from axioms. Hilbert envisioned his second problem as the demonstration that the results of the two would always be equivalent; that is, it should be possible to logically deny or confirm any grammatically well-formed statement, just as it should be possible to construct grammatically any logically proven statement. When Gödel constructed the proof for the incompleteness theorem, he did not choose logic to derive a statement. Instead, he freely created grammatically correct statements and showed that some of these statements cannot be derived logically using axioms. Since the statements were freely created, it was possible to embed into them empirical and mental criteria, while remaining within mathematics. This undermines the view that mathematics is free of mental meanings or empirical truths. Freely created statements can bring in mind and matter by using distinctions from ordinary language.

Issues with Hilbert's program are also apparent when we examine the nature of symbols, which mathematicians of the time did not consider an important issue. It turns out that to even define symbols we must distinguish them using physical and conceptual means. How else do we tell one symbol apart from another? An ordinary symbol like 'A' or 'B' is distinguished using a physical property—their *shape*. Of course, shape alone is not enough to distinguish the letters because the same letter can be depicted using many different fonts. Aside from the physical property of shape, therefore, we must also learn to distinguish the symbols using types. These types are understood through the interrelations between them. Words similarly must be distinguished by both their physical shape and their mental meaning. Physical and semantic considerations must therefore enter mathematics, even in the simple exercise of trying to distinguish symbols, and the Formalist reduction of mathematics to symbol manipulation does not liberate us from distinctions between conceptual and physical. Physical references serve to validate the truth of a statement if the object referenced in the statement *exists*. However, everything meaningful need not exist. To claim that we *understand* what it means, we must allow for a separate mode of existence when something is meaningful but does not exist versus when something is meaningful because it exists.

Philosophers in medieval times had believed that everything that was meaningful was also true. Thus, they relied on showing that the idea of God is *meaningful* to conclude that it is also true. Medieval philosophers developed intricate arguments from first principles to show that the idea of God is meaningful in order to conclude that God must exist. After all, how can something that is so meaningful and is based on logic not exist? Is the universe not meaningful? String Theorists in physics similarly claim that their theory is meaningful to conclude that it must also be true. The meaningfulness of String Theory makes its truth likelihood higher, in so far as we believe that the universe is meaningful. Frege wanted to escape this conclusion, and he argued that everything meaningful is not necessarily true. But his conclusions now also applied to mathematics! In short, not every mathematical statement need be provable or disprovable.

Russell's Paradox

An indirect way of solving Hilbert's second problem is to reduce all of mathematics to logic. The three principles of logic—identity (A is A), non-contradiction (nothing is both A and not-A) and excluded middle (everything is either A or not-A)—are definitions of consistency and if mathematics can be derived from logic then it will be consistent by default, because logic is consistent by definition.

Of course, it is debatable whether logic alone is sufficient to derive all of mathematics since, traditionally, mathematicians have used logic together with axioms to create knowledge. Euclid's geometry was for example a product of the application of logic to axioms. Axioms are self-evident ideas, which seem to be intuitively true although never actually provable within the logical system. Since it is hard to even derive these axioms from more fundamental principles, axioms are expected to be irreducible assumptions upon which a mathematical theory is based. Hilbert wanted to identify a method by which any set of axioms can be made non-overlapping and consistent. What better way than to reduce the axioms to logic itself? For, if the complete set of axioms is derived from logic then they ought to be consistent because logic itself is (supposedly) consistent[4]. The idea that a complete set of axioms can be derived from logic is rationalistic to the hilt as it assumes that everything knowable is rational. Logicians found it tempting to construct axioms from logic, as the route to proving their consistency and completeness.

Gottlob Frege, Bertrand Russell and Alfred North Whitehead set themselves the task to create a foundation for mathematics based on logic. It was well-nigh impossible to create all of mathematics just from logic, so something had to be added to logic to create "mathematical logic" as a foundation for mathematics. The construct adopted was that of a "set" as a collection of objects. The original idea behind choosing a set as an elementary mathematical construct is that it can be used to represent any arbitrary *concept*. Russell believed that a concept is known not by grasping a mental entity that we might call its meaning but by expanding the concept into an elaboration of the things to which the concept applies. For instance, the meaning of the concept *car* is the collection of all cars, to which the concept can be applied. Any concept can be represented by its *extension*: the collection of things to which the concept can be applied. Here lies the germ of a problem that I will discuss later—that a concept is defined not just by its *extension* but also by its *intension* or meaning. However, since there had been no good way to talk about meanings within mathematics, concepts were represented by their extensions or sets. The idea began to prove useful when Russell showed numbers can be represented by sets. The representation only required the assumption that a number is a concept and every concept can be represented by its extension using a set.

Russell treated the number 2 as the generalization of the concept of twoness. Accordingly, the concept is seen in every pair of objects— such as the collection of two shoes, two cars, two people, etc. Thus, the number 2, Russell said, could be represented as a set comprised of sets of two shoes, two cars, two people, etc. As you can imagine, there are infinitely many sets of two things, and the concept 2 denoted as a set has an infinite number of members. Cognitively, this scheme is not representative of how we learn the concept 2. For example, we don't necessarily know about every possible pair of objects before we understand the meaning 2. In fact, given the practical difficulty in knowing every possible pair of things in the universe, it seems impossible that we would ever understand the number 2 if we had to know about every possible pair in the universe. But Russell was not creating a cognitive theory of numbers, of how we understand and learn the idea of numbers. He was only interested in a logically consistent foundation of mathematics. As we will shortly see, Russell's attempt did not work, and this means that the question of what numbers are is still open and it perhaps needs an idea more representative of how we understand numbers, different from how their definition can be provided solely based on logic.

There is another problem with Russell's definition, namely that to know that a set does indeed have 2 shoes, we must have the concept of two even prior to counting. How else do we know that there are 2 and not 3, 4 or 5 shoes? Our assessment that there are 2 shoes depends on our ability to distinguish objects and count them. But to count shoes, we need the numbers which we are trying to define, and the definition cannot assume the existence of something being defined; it can assume things that are previously defined. How can we assume counting when we have set ourselves the task to define it?

In any case, with this definition, it became possible to reduce a number to logic and sets, paving the way to the eventual reduction of mathematics to logic and sets. Russell and Whitehead used sets with rules of logic to prove the truth of statements like "1 + 1 = 2".

Now, it was natural to ask: If a collection of objects is a set, then what is a collection of sets? Undoubtedly, the answer had to be that a collection of sets is also a set for this notion to be powerful and useful enough in mathematics to be used as mathematical foundations. A set was thus defined to be a collection of any kind of object, including sets of objects, the sets of sets of objects, and so forth. The need to treat sets like objects (members of other sets) introduced yet another problem for counting. We can talk about the 'set of all sets' and since this is also a set, it must be a member of the 'set of all sets'. The problem now is that a set is a member of itself. As a set, it denotes a concept, but as a member it represents an object. The old dichotomy between concepts and individuals returns to haunt us. This problem quickly led to a paradox that is now called *Russell's paradox.*

Russell constructed a paradox with such self-including sets. He defined a set *S* as the set of all sets. He then asked: Is *S* a regular set or a self-including set? If we say that *S* is a regular set then by definition, *S* must belong to *S* since *S* is a set of all regular sets. This makes *S* self-including and leads to a logical contradiction with the assumption we started out with—namely that *S* is a regular set. If however we say that *S* is a self-including set, then we have a bigger issue in constructing *S* because of infinitely recursive inclusion. In the former case, there is a contradiction; in the latter case there is incompleteness because we can never construct the set *S*.

Russell also invented a way to solve the problem. He said that a set is not an arbitrary but a *class* of homogeneous collections. It is possible to collect two shoes, three shirts and five ties in such a set, but not include a set within itself. A collection of shoes, shirts and ties can represent a concept such as 'male apparel' but a set within itself denotes nothing real. It is thus possible to associate a concept with every kind of collection except those that include themselves.

Russell claimed that a set can include only objects of the same *type.* In the simplest case, a set is a collection of objects. Next, a set can contain sets that contain objects. This could be followed by sets that contain sets which in turn contain sets that contain objects. At every stage of cascading abstractions, a set will also represent a *class*, containing objects of one type. Sets that contain themselves may be mathematically possible, but these do not represent anything meaningful, and we can leave them out of our theory. The set that collects all regular sets is therefore left out in this theory.

Now, this may address the specific contradiction that Russell discovered, but it doesn't solve the deeper problem—namely how we form sets. To make a set of horses, we must have the concept of the horse—i.e. the *intension*—before we can identify things as horses into a collection. Similarly, to form a set of five members, we must have the concept of five even prior. Therefore, if you have the *intension* then you can construct the *extension* or the set. But if you just have objects, you cannot construct the extension because you won't even know how many things there are let alone the types of things. But Russell was adamant. He formed the Axiom of Reducibility in which the lowest level in the lowest level entity is an indivisible object without any parts. All subsequent collections of objects and collection of collections create levels of conceptual hierarchy. Russell called this ordering of objects and sets the *theory of types*. The idea was that each set represents a type and only objects of that type can belong to that set. Sets that include themselves fall into no clear type and the case where objects and collections belong into the same set is unrealistic and does not represent any conceptual entity. In trying to solve the paradox he discovered, Russell took a path in which we did not have to entertain concepts *a priori*. He wanted to construct all concepts from objects, but as we can see, it is impossible to form sets without acknowledging the prior existence of concepts.

The Axiom of Reducibility states that all concepts can be reduced to objects. This was a continuation of the thesis that statements are meaningful only when they can be verified. If all concepts are formed based on collections of things, then we can say that all concepts are derived from the external world. The problem, however, is that even to collect things we must *classify* them using *a priori* concepts. The problem that Russell and others had set out to solve—namely, resolve the contradictions in mathematics by reducing mathematics to logic—failed because set theory, which was supposed to salvage the situation in number theory turned out to be riddled with issues.

Gödel's Incompleteness

While the Logicist program to reduce mathematics to logic did not work in all respects, Hilbert's original problem still seemed relevant. Reducing mathematics to logic was one way of achieving Hilbert's dream, but with the failure of the Logicist program it now became necessary to identify alternate approaches to achieve the same goal. Hilbert had envisioned that the problem would be solved if we could identify a definite procedure that could prove if a collection of axioms was non-overlapping and complete. Using this procedure we would evaluate different axiom sets until we find the right axiom set that could provide a solid foundation for mathematics. With the Logicist program failing to impress, identifying a procedure to determine internal consistency amongst axioms still seemed like a valid problem to attack. The problem had however to be worked on under the premise that mathematics will have to use both self-evident axioms and rules of logic (logic alone was insufficient). The proof would need us to show consistency and completeness by using both logic and axioms and not just logic as the Logicists had been trying.

This is where Gödel's incompleteness theorems are important. Gödel proved that no proof scheme can ever be used to demonstrate both the consistency and the completeness of axioms. But he did not just mean that the procedure for determining consistency does not exist. He went on to say that the procedure does not exist because a consistent and complete axiom system itself does not exist. These two are distinct claims. We can claim that there is no automated procedure to decide if an axiom system is consistent and complete, without saying that there is no consistent and complete axiom system. The latter claim is stronger and Gödel made the stronger claim. Gödel's theorem means that every axiom system is incomplete and every axiom system will have unprovable statements. There is no automated procedure to prove consistency and completeness because every axiom system is either inconsistent or incomplete.

Gödel proved two incompleteness theorems and they pertain to different aspects of the problem. The first theorem (let's call it T1) proves that no axiomatic system will ever be complete and every axiomatic system will have statements that are meaningful although they cannot be proved. The second theorem (let's call it T2) proves that no axiomatic system can be proved to be consistent from within the axiomatic system. If we embed an axiomatic system within a larger set of axioms, then by virtue of T1, the new system of axioms will have unprovable statements and the new system's consistency cannot be proved because of T1. These two theorems destroyed Hilbert's dream of complete and consistent mathematics.

T1 says that no axiomatic system that deals with numbers can be both consistent and complete. Mathematicians prefer consistency to completeness. It is better to not prove some meaningful statements than to prove some false statements! Gödel also proved that logic itself is consistent. So his proof is taken to be a denial of the completeness of number theory rather than a proof of the inconsistency of number theory. Since mathematical knowledge grows through proofs, the inability to prove some meanings means there can be true conjectures in mathematics which we cannot prove.

T2 says it is impossible to prove the consistency of an axiomatic system from within the system. We may attempt to prove the consistency of an axiom system by embedding one axiom system into another larger axiom system. But the new axiom system is incomplete by virtue of T1, and unprovable statements in that larger axiom system may include statements formulated in the smaller axiom system. The "proof" of the consistency of the original axiomatic system, within the larger axiom system, is now subject to both inconsistency and incompleteness. That is, the proof may not be possible because it is an unprovable statement, and even if we did find a proof, we can't be sure of its truth due to inconsistency.

Gödel's theorems imply limits to how much reason can be used as a method to knowledge. Traditionally, a logical system is more versatile if it has a greater breadth of axioms. Now, incompleteness theorems imply that no matter how numerous the axioms, there will always be some truths that cannot be reached from the axioms.

To imagine this visually, think of a planet with oceans and continents. Assume that the axioms in a theory are some of the islands. Islands that can be reached from the axiom islands through a land or water route represent theorems that can be proved using the axioms. We would generally assume that every island on the planet is reachable from the axiom island, but Gödel's theorem says that there are some islands that can never be reached because there are no routes connecting these islands. Although we believe that these islands exist, we cannot prove that these islands truly exist because we can never conclusively go to the place where the island purportedly exists to confirm or deny its existence. Our knowledge is restricted to the islands that we inhabit or can reach. Gödel's theorem says that every axiom system will have unprovable statements and that axiom systems are like planets that will always have unreachable islands. Attempts to reach such islands will be futile. But, since we cannot know which attempts pertain to unreachable islands, there is no way to prevent the attempts upfront. Attempts to prove statements on disconnected surfaces must end up drowning in the sea or returning back home safely although unsuccessfully.

Gödel's theorems make the practice of mathematics harder than previously envisioned. With modern technology, it is possible to imagine that computers can prove theorems automatically, and algorithms for such automated proving could be defined. But the existence of such an algorithm also depends upon the assumption that all mathematical propositions are either provable or disprovable. Gödel's theorems imply that there is no way of knowing upfront if a theorem is unprovable. If we are trying to prove (or disprove) a proposition, and haven't found the proof (or contradiction) yet, we cannot know whether to expend more effort to find the proof or ignore the problem because it can never be proved. A mathematician may spend his life trying to prove a proposition, when, in fact, the proposition is unprovable. It would be a life lived in vain, although we cannot know it in advance unless a problem is shown to be unprovable (which is possible although not always easy). To crack a hard problem, we should therefore alternate between proving the problem or showing it is unprovable.

Gödel's theorems cast a shroud of doubt on all unsolved problems. Given that a problem hasn't been solved, should we assume that it is in principle unsolvable and therefore not attempt solving it, or is it that we just haven't found a solution for it and the problem is in principle solvable? Since Gödel's theorem does not say which problems are unsolvable, it is difficult to tell upfront if a problem is ultimately unsolvable. And we continue trying to solve problems in the hope that they are ultimately solvable. Some of these attempts, even if the theorems are indeed true, could be futile.

Turing's Halting Problem

While Gödel had showed that number theory would be incomplete, there wasn't yet a clear example of a specific kind of problem that is undecidable. Alan Turing supplied the first instance of such a problem[5]. Turing was a pioneer in computer science and he formalized the notion of an algorithm through his Turing Machine that reads instructions and modifies inputs into outputs. An algorithm is considered finished when the Turing Machine comes to a halt. However, it is possible that a problem is unsolvable and then the Turing Machine would continue indefinitely. Since there could be complex algorithms that take a long time to finish, it is important to determine whether an algorithm will ever halt. There is no point in expending a computer's energy in trying to solve a problem that is unsolvable since no amount of computation could ever prove or disprove its statements. The question of whether an algorithm will stop is called the Halting Problem and it indicates whether a machine that tries to compute a solution will ever come to a halt.

Turing proved that the Halting Problem is an undecidable problem, or an example of Gödel's unsolvable problems within arithmetic. Turing's proof means that there is no procedure to decide if a program will meet its intended goal and stop. The only way to know if a program will terminate is to run it on the computer and find out if the program terminates. The essence of the Turing's limitation can be illustrated by the following simple program:

INSTRUCTION 1: GO TO INSTRUCTION 2
INSTRUCTION 2: GO TO INSTRUCTION 1

The casual examination of these instructions shows an infinite loop between them. The first instruction asks the computer to go to the second instruction which puts the execution back on the first instruction. A computer can never exit this infinite loop because there is no instruction that tells it to stop. If you run this program on a computer, the program will run forever, and it is advisable that we never start a program that is never going to conclude. Turing's proof of the Halting Problem means that there are no formal procedures to distinguish programs that halt from those that don't.

This illustrates the contrast between computer programs and humans. Even an average intelligence human is unlikely to loop through the above instructions more than once. Humans would quickly detect a loop and stop even though there is no instruction to that effect. Humans are goal oriented and can see that looping is not taking them closer to the goal of solving a problem. A computer is not goal-oriented and has no way of knowing if it is getting closer to its goal. It knows how to execute instructions but has no clue about the computational 'distance' between a problem and its solution. When faced with an intractable problem, a computer would continue indefinitely on a line of approach that has been fed into it through programming. Human beings will likely alter their approach, try to solve the problem from multiple angles, and take the ideas and intuitions developed in one approach into another. They might bring unrelated ideas to bear upon the solution of a problem, which a computer will not. In case the problem isn't solved, humans would stop attempting after a while, but the computer will not.

In short, computers can never stop even when the problem is unsolvable and Turing formalized this in the Halting Problem. A problem might take a hundred years to solve, so it is worthwhile to know that the problem indeed has a solution before we spend a hundred years trying to solve it. It would be futile to spend a hundred years and then abort the attempt because the solution wasn't found so far. Humans have the ability to abort intractable problems and Turing proved that this was impossible for a computer. The Halting Problem is an example of the kinds of unsolvable problems that Gödel's theorem alludes to, but did not explicitly identify. The machine that attempts to answer such a question for a program that never halts will also run forever since coming to a stop means determining that the program being analyzed also comes to a halt.

Fortunately, Gödel's theorem does not prevent us from identifying unsolvable problems. The theorem just says that there is no *universal procedure* to identify all unsolvable problems. But it may be possible to identify techniques unique to determining whether a specific problem is undecidable. Turing's proof of the Halting Problem is an example of such a technique. The Halting Problem is not unsolvable according to Gödel's incompleteness theorem but it was proven unsolvable by Turing's proof. Other unsolved problems may be proven undecidable in a similar fashion. A computer that has been unsuccessful in proving a theorem could still prove that the problem is unsolvable. The proof of unprovability however cannot be obtained through a predetermined procedure. The advantage of such proofs is that we will not try to solve a problem that has been shown to be unsolvable because such a machine is clearly not possible. It saves a lot of effort, although the outcome is hardly consoling. We might just find solace in that we failed without expending the effort in futile attempts instead of failing after much endeavor.

If Gödel's theorem puts a limit on knowledge by asserting that some meaningful statements cannot be proved, the Halting Problem hammers a wedge between man and machine. It says that if a machine is in an infinite loop, it cannot stop itself, because it cannot know that it will never end. If, however, the machine stops, the credit for it must go to the programmer who provided the operating instructions, since the computer itself is dumb. A machine can follow instructions but the computer programmer must provide the correct instructions. The Halting Problem also implies that a computer cannot create algorithms to solve a problem, because a computer cannot judge if an algorithm will solve a given problem. This means that a computer cannot be given problems to solve, but only procedures that, when applied, will solve the problem. The computer can blindly follow the procedure since it has no independent method to decide if the procedure is actually fit for the given problem.

We can conclude that the "process" by which humans analyze, understand, diagnose, and formulate solutions to problems cannot be encoded in a computer. For, if this were possible, then a computer would know which programs will solve a problem, and hence know if a solution meets the goals of a problem. Humans must therefore design algorithms for solving problems, using their creativity and ingenuity, which a computer can execute faster than a human[6].

This puts serious limitations on a machine's ability to solve problems. A computer is useful if an algorithm has been defined. The computer however cannot do the necessary thinking needed to arrive at a suitable algorithm—a computer cannot be goal-oriented. We can only program procedures previously invented and not problems whose solution methods are unknown. Since the mapping between a problem and its solution lies with the programmer, the computer has no clue what problem it is solving when it executes instructions. If an algorithm takes its execution away from a problem, humans will recognize that the solution isn't appropriate for a given problem but a computer will not. Humans will stop their programs, change the solution method and start all over again. A computer will continue executing the same program whether or not it really works. A computer can look smarter than a human being due to its execution speed, although without the algorithms that go into it via the software, the computer is dumb. Turing's proof of the Halting Problem asserts that all Turing Machines will always be dumb.

This fact about Turing Machines helps us appreciate the nature of incompleteness and how it applies to machines. Incompleteness applies to humans as well, in so far as we have machine behavior, for example, in theorem proving using algorithmic means. Of course, humans don't always use algorithms. Humans have the ability to step out of logic and look at the act of logical thinking in perspective. This ability to logically think about logical thinking is a meta-action (action about other actions), an ability that machines do not possess. When tasked with a problem, humans *plan* by laying out the steps to solve the problem before they actually start executing those steps. Many alternatives may be considered during this planning, and some alternatives might be discarded by applying constraints. By the time the planning is complete, there is a clear path connecting the start to the end, and we know that if we were to execute the plan correctly we would attain the solution. Of course, the plan might not always work. The plan may be faulty and it might leap from one state to another without making a connection between them. When such things happen, they are treated as further problems to be solved requiring another plan. Human problem solving thus alternates between planning and plan execution and we first theoretically solve a problem by a procedure before we practically execute the procedure determined as a plan. Computers are however incapable of planning, although they can execute a plan very well.

Turing's theorem illustrates some basic differences between human thinking and machine processing. Humans can think about program semantics (i.e. what a program does) while computers can only execute instructions. Many people believe that this is a curious property of minds which could not be embodied into machines. I would instead argue that this is not actually a property of minds, but that of ordinary language and the distinctions it uses. Computer programmers use ordinary language to communicate program semantics. The Halting Problem pertains to a distinction between how we use ordinary language versus how we use mathematics. Unless we suppose that ordinary language is mental and mathematics is not, we can imagine that it would be possible to solve problems of incompleteness by incorporating missing features of ordinary language within mathematics. The Halting Problem can be solved by incorporating program semantics in computing theory.

Modalities in Language

Ordinary language involves the use of several distinct modalities, which can be denoted by the same word. The context of usage in ordinary language tells us the mode to which the word presently refers. In mathematics, however, the mode associated with a number cannot be decided contextually, although the modes inherent in ordinary language can be present in mathematics as well.

This lies at the root of paradoxes in logic, mathematics, linguistics and computing, which use ordinary language modes but cannot distinguish between them. Historically, philosophers have debated what a word truly refers to. Does a word refer to sensations? Or to things that exist 'behind' those sensations? Does a word actually refer to some concept or do words not refer to anything at all (i.e. words are just sounds to which we mentally attach some concept?). These debates have raged for a long time in the philosophy of language but haven't received much importance in mathematics. My contention is that Gödel's paradox is an outcome of the inability to clearly differentiate between different uses of a word. In a sense, the paradox illustrates what has never been settled adequately in mathematics—namely that all the ways in which words are used in ordinary languages are somehow equally real. The word therefore does not have a single fixed reference. Rather, language has many distinct kinds of *models* to which it can be equally applied.

The Real Mountain   The Concept Mountain   Picture of Mountain

Figure-1 Ordinary Language Categories

To better understand these categories consider the different meanings of the word 'mountain'. The word 'mountain' can denote a real mountain, a concept mountain and a picture of a mountain. The real mountain is a thing, and each thing has an individual identity. The word 'mountain' can refer to that identity. The word can also represent a concept, which is a class of things, not a specific thing. Of course, concepts are generally derived by abstracting the properties of real things. Once abstracted, the concept can be applied even to things from which they would not have originally been abstracted. Thus, for instance, we can speak about a mountain of papers or a mountain of garbage, even though, strictly speaking, they are not an individual thing or identity in the sense that we think of mountains as physical things. We would also not likely have derived the idea of a mountain just by looking at a pile of garbage or paper. Context sensitive association allows us to apply concepts to things in novel ways, and the association between a concept and a thing is not rigidly defined. In calling something a mountain of paper, we somehow know that we are using the word 'mountain' as a concept and not referring to a real mountain.

Of key importance here is the fact that we distinguish between a thing and the concept, and while the thing is physical, the concept is supposed to be mental. Both the real mountain and the concept mountain are not *referential*. That is, they are physical and mental objects, but they don't necessarily *point* to each other. For instance, there can be a large object which is not necessarily called a mountain. Similarly, there can be concepts which don't have any counterpart in the external physical world.

The physical and conceptual are merged in a *symbol*. It is important to note, however, that there are two kinds of symbols. First, there are symbols that *denote* a concept. Second, there are symbols that *describe* an object. Street signs, for example, denote a concept, such as stop, pause and go. A book or a picture, on the other hand, generally describes facts about the world. Of course, this distinction is not rigid and definitive. For instance, a picture or a book may not describe any facts about the world; abstract art and fictional books are representations of concepts, and not descriptions of reality. Similarly, some street signs may describe that the road curves ahead or lanes merge ahead. Nevertheless, the distinction between denotation and description is pervasive although when a symbol denotes versus when it describes depends on the context. A picture of a mountain is a physical thing, but it represents or points to a real mountain. Unlike the real mountain, which can be in some situations treated as a denotative sign of the concept mountain, the picture of the mountain is a description of the mountain but a denotation of art. Of course, one can view the symbol entirely as a physical thing without a meaning; this happens, for instance, if one is not familiar with the descriptive language in which the symbol is encoded or when the contextual denotative significance is not seen.

These three ways of interpreting a word are also philosophically present in mathematics as different ideologies about mathematics. The Realist views mathematics as the description of the real world. The Platonist thinks of mathematics purely as ideas. The Formalist sees mathematics purely as tokens which can be encoded in machines and matter and can be used in theorem computation. In fact, Gödel successfully incorporated all these three types of meanings within the same sentence. The distinction between things, ideas and pictures that exists in ordinary language was dissolved in Gödel's proof because mathematics cannot distinguish between the different meanings of the same number. Before we see how Gödel incorporated three different kinds of categories in the same proposition, let us see how something like this could be achieved in ordinary language. Consider three different meanings of 'mountain':

1. The word 'mountain' has eight letters.

2. A mountain is very high.

3. That mountain is very high.

In the first sentence, 'mountain' denotes the word as a physical thing. The thing does not itself denote the meaning unless we know the linguistic convention being used to interpret it. In that sense, the word 'mountain' has a meaning, but if we don't know the language then it is just a thing. The second sentence uses 'mountain' as a concept, describing the general property of mountains to be high. If there were no mountains in sight or no one had ever seen a mountain, the word will denote a mental concept. The third sentence uses the world 'mountain' as a *name* for something the sentence refers to or describes. The above examples illustrate how the same word can be viewed as an object, as a concept and as a referential name.

Gödel's trick was to interchangeably treat a number in three different ways. He formed sentences of the kind P: P is false. The first P is a thing that represents the sentence. The second P has two interpretations. First, we can treat it as a name by which a sentence is called. Second, we can treat P as a meaning of the name.

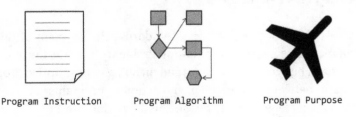

Program Instruction   Program Algorithm   Program Purpose

Figure-2 Three Interpretations of a Program

Of course, names, concepts and things are not the only categories in nature. Computing theory specifically employs another broad category—*actions*. Just as a word can be interpreted as a thing, a concept and a name, similarly, a program can also be instructions, algorithms and problems. Program instructions are analogous to physical things. Every object has some physical properties which can be described. But every object also has an ability to cause a change, which is known through the causal relation between the property and its effect. The same object can therefore be interpreted as properties as well as effects; the property may be mass and its effect may be motion. The same effect may be produced by a different property, and the same property may produce a different effect. In that respect, properties and effects are not identical, although every property will also have a corresponding effect. Properties and effects are two different categories in ordinary language, but Turing's Halting Problem uses them interchangeably. Turing treats a program both as a set of physical properties (data) and as a set of effects (instructions). Data can also be denoting concepts and names, as we have seen previously. Similarly, instructions can also be interpreted as program algorithms and the problems that it can solve.

Program algorithms are analogous to descriptive concepts. And program purposes are analogous to descriptive references or names. Like a statement describes reality, similarly, a program solves a problem. Like Gödel's theorem exploited the confusion between names, concepts and things, Turing's theorem exploits the ability to dually-interpret a program both as a thing and as an action.

Solving the Paradoxes

I have so far described one way to address the problem of Gödel's Incompleteness and Turing's Halting Problem—namely, the ability to distinguish between name, thing and meaning in descriptive propositions and between name, thing, and action in programs. This is how ordinary language does it, but in this book, I'm not going to advocate this solution. Instead of finding ways to contextually differentiate between different modalities in language, I would suggest that we allow the equivalence of names, things, meanings, and actions, but disallow the *inconsistencies* that arise because of this interchangeability. For example, I would suggest that we prevent the formation of sentences of the type P: P is false. The ability to call "P is false" as P exists in current mathematics, because the *name* P does not indicate the *meaning* P. Thus, we can call a white rose as a black rose. When a name is equated with the meaning, it leads to the conclusion that black is white. For mathematics to prevent such occurrences, we must fix the way symbols are given identified consistent with their meanings. The symbol is the 'thing' in question; its identification or distinction from other symbols is the 'name' we attribute to it; and this name must represent its 'meaning'.

I also want to ground this solution realistically in terms of a theory of objects in space and time. In the everyday world, objects are identified by their location in space, and the location in space and time represents a unique *name* by which the object can be distinguished. The object itself is in this case the symbol, and its name is its location in space and time. However, this is not enough to solve the problem at hand, because any object can be placed at any location in physical space, and that would mean that any symbol can be called by any name, which would then lead to the above contradictions. To truly address this issue, we must change the nature of space from *physical* to *semantic*. What is semantic space? It is a space in which location represents *meaning*. To be at a different place in this space entails to denote a different meaning. The location is also the name, so by this equation of names and meanings represented by a thing, we avoid self-contradictions arising in the above paradoxes.

Truly speaking, this is not a space of physical objects; it is rather a space of symbols in which the location represents meaning. Therefore, there can never be a situation in which an object's meaning is P but its name is not-P. Gödel's Incompleteness, Turing's Halting Problem and other paradoxes arise in mathematics because space is used to distinguish objects but not their meanings. When the location is used as a name and the name is then interpreted as a meaning, this leads to a contradiction because the object with the name not-P has a meaning of the concept P. The solution to these problems is to dissolve the distinction between *naming* and *meaning*. This requires us to formulate a natural language in which the symbol 'apple' is situated at the location 'apple' and means 'apple'. This approach allows us to consistently interpret a symbol as a name, concept and thing without contradictions. The approach is Realist, because it deals in space-time objects. It is Platonic because it deals in ideas. And it is Formalist because it deals in symbols.

Once this general approach to semantic space is recognized, we must extend it for programs in addition to descriptions. The extension requires us to acknowledge the existence of another *dimension* that represents the *actions* of the symbol. Colloquially, we can view them as nouns and verbs; the nouns are the descriptive meanings and the verbs are the action meanings. Every symbol represents some meaning which can be understood in two ways—through perception of what it *is* and what it *does*. For example, we can learn about a car by observing its shape, size, color, and the analysis of its parts. Each of these constitute a *description*. However, the thing that looks like a car isn't necessarily a car unless it works like one. Therefore, the observation of its shape, size, color, and parts is incomplete, unless we complement it with its actions or effects. Likewise, the portrayal of actions is also incomplete; something that runs on four wheels isn't necessarily a car; it can also be a truck or a bus. Therefore, the actions of a thing also incompletely identify the object. However, the *combination* of the two complementary aspects constitutes a complete definition of what that object is.

Therefore, when we speak about a semantic space, we primarily need to think of it in terms of two dimensions—a descriptive and a prescriptive. The descriptive dimension denotes how we perceive that object itself, and the prescriptive dimension denotes the effects it has on other objects. The *name* of the object is then the position of the symbol in this two-dimensional semantic space. And since this location represents meaning, it also denotes the meaning.

Now a few words must be said about how location and distance in space can represent meanings. The fundamental fact in relation to meanings is that they are defined in relation to more *abstract* meanings. For example, the idea 'dog' is defined in relation to the idea 'mammal', and the idea 'mammal' is defined in relation to the idea 'animal'. Similarly, a 'sedan' is defined in relation to 'car', which is defined in relation to 'vehicle'. These relationships always construct an *inverted tree* geometry, with the most abstract idea forming the root of the tree at the top, with less abstract ideas forming its trunks, branches, and leaves. The less abstract ideas inherit the properties of the more abstract ideas, but not vice versa. In that sense, there is a strict sense of direction in the tree—from root to leaves.

Once we see that the concepts are defined in relation to other concepts in an inverted tree, we must define our semantic space too as an inverted tree, rather than the flat Euclidean (nor curved non-Euclidean) space in modern physical theories. The key distinguishing character of the physical space is that it has finite number of dimensions; e.g., space in Newton's physics has three directions. The semantic space on the other hand has literally infinite dimensions; each trunk, branch, or leaf in the tree is a dimension; so, if the tree has infinite diversity then space too is infinite dimensional. And yet, we can represent this entire tree in a three-dimensional container *as if* it had only three dimensions. Collectively, the tree is present in a three-dimensional space, and yet it has infinite dimensions. Thus, we don't have to reconceive an infinite dimensional space to understand semantic space; we just must reconceive the *geometry* of this three-dimensional space to be an infinite-dimensional structure.

Since the higher rungs of this inverted tree are more abstract than the lower rungs, it is impossible to *physically* identify these higher rungs. This is quite like the fact that we cannot identify a separate entity called the 'mammal' if we only observe the dog. The 'mammal' is both *inside* each dog, and yet *outside* of them. The mammal is inside each dog because the properties of being a mammal are present in each dog. And yet the mammal is outside the dog because that property exists in other dogs, or even cats. The idea 'mammal' is therefore not *physically contained* in any dog, and yet it is present by the properties of the mammal in the dog. This is a curious property of the semantic space, which cannot be understood if the geometry is *flat*—i.e. has a finite number of dimensions. It is easily understood if the geometry is *hierarchical*—i.e. as an inverted tree. Therefore, the fact that the mammal is both inside and outside the dog is inconceivable in the physical space, but it is eminently conceivable if we reconceive the geometry hierarchically.

In fact, we can now give a very precise definition to the three dimensions in which the tree exists. We can think of this inverted-tree like space as a layer cake—the layers broadening at the bottom of the cake and narrowing toward the top. Each layer is two dimensions of description and action, and the successively higher layers represent more abstract ideas of description and action in relation to which the detailed ideas (the lower layers) are defined.

We can also give a precise definition of the notion of *position* in this tree-like space—it is always defined in relation to the root of the tree, through the successive trunks, branches, and leaves. The 'distance' between two trunks, branches, or leaves is the *path* that takes one from one part of the tree to another—by traversing the graph structure, and without stepping outside the tree itself. For instance, the distance between two leaves of the tree isn't the direct straight-line path between them even though they may seem to be very close; it is rather the path that joins the leaves to the nearest branch and the distance is from one leaf to the branch and back to the second leaf—rather than the straight-line path between them.

The nodes of this inverted tree graph structure are the *symbols*. The edges of this graph are the distance and the meaning. The distance helps us define the *location* in this tree, and that is identical to the *name* by which we can call the symbol. Since that name is identical to the meaning (because the nodes of the tree are concepts), the form of the symbol becomes the thing that symbolizes a meaning through a name. We still have the same three modalities we spoke about earlier—namely, meaning, name, and thing—so we haven't taken away the power of language to operate in these modes. However, we have taken away the cause of contradictions.

The source of contradictions can now be traced back to the use of flat or finite-dimensional space in which any symbol could be placed at any location, thereby giving it arbitrary names, which entails that the name by which we call that symbol, and what it means linguistically, are detached. Once name, meaning, and thing are detached in the flat space, then contradictions are imminent. You can now place the proposition P in the location called not-P. Since you would like to interchangeably use naming and meaning, you must now introduce separate modalities to avoid the contradiction because the fact is that names and meaning are not identical. And yet, the use of modalities complicates the language immensely because there are numerous interpretations of a sentence depending on which mode is applied to which word. One such modal interpretation cannot be used to infer conclusions in another interpretation. But, how do you consistently apply these modes across many sentences? How can you be sure that you are not going to create contradictions as you evaluate a collection of modally interpreted sentences? That's the problem that the semantic space approach resolves. It basically says that you can use any mode to interpret any sentence and you will never obtain a contradiction because we have devised a unique language in which names, meanings, and symbols are identical.

Understanding Natural Language

Now you might argue that this unique language that we have devised for mathematics isn't the personalized and contextualized natural language that we employ in daily use, and you would be correct. The first step in understanding natural language is not natural language itself; it is making mathematics a language of concepts (i.e. give it the ability to employ modalities) free of the problem of incompleteness. This creates a complete and consistent *universalist* language. It is still not the personalist or contextualized natural language. However, if we have addressed the issues of the universalist language, we now have the tools to address the problems of natural language also.

The solution is to extend the universalist semantic space to cover the problems of personalist and contextualist languages as well by postulating two other spaces—personal and contextual—that are like universalist semantic space (i.e. they have an inverted tree like structure) although different in one key sense: the distances between symbols (and hence their meanings) have personalized meanings (in the personal semantic space) or cultural meanings (in the contextual semantic space). This allows arbitrary words to have arbitrary meanings, but in fact they are not arbitrary. They are just personal and cultural associations between words and meanings that form personal and contextual semantic trees. There is just one universalist language, but there can be numerous personal and cultural languages. Indeed, it is not only possible, but also likely that the imprints of the universalist language are present in many if not all personal and cultural languages, and these languages may share similarities with the universalist language to different extents.

In the universalist language, the distinction between name, meaning, and thing is not required, but it becomes necessary in the personal and cultural spaces. This naturally means that machines cannot process the personal and cultural languages correctly unless they embody the same personal and cultural structure embedded in the users of the language. Even if they do, they would have to be as many personal and cultural structures as there are persons and cultures. No universalist scheme would always work correctly.

Humans are adept at processing personal and cultural languages because they incorporate the personal and cultural spaces in addition to the universalist space. The universalist space is partially represented in our scientific, logical, and mathematical instincts where we try to think of the world in universal ways. But aside from these instincts we also have a personal and cultural space in which we must distinguish between the modalities in language because the names, meanings, and things are not encoded identically.

Humans are well-equipped to distinguish between the modes being used in a context sensitive manner. For instance, we can know when a certain word like 'address' denotes a noun or a verb. We can also tell naturally when the word 'barber' represents an individual person or a general concept. We draw these inferences not just by looking at a specific sentence, but by trying to reconcile a lot of sentences together in a given context, for a given person, in a given culture. In short, we *assume* that the conversation is rational, so it must be *consistent*. We can *choose* an interpretation of the sentences that makes the conversation rational and consistent. In short, we are not just deriving the truth from *a priori* given meanings. We are also doing the reverse—we are assuming that there is truth and consistency and we are using that to derive the meaning.

There is a complex bidirectional interplay between meaning and truth that occurs in humans, which doesn't occur in mechanical symbolic systems that can be automated in current machines. Machines operate unidirectionally—from meaning to truth. The machine takes a universal dictionary of meanings for granted, uses it to derive the meanings of sentences, and it must operate in a specific modality based on the dictionary it has been supplied. For instance, the dictionary must always treat 'address' as either a noun or a verb. Once this dictionary has been adopted, the meaning is determined, and then logic can be used to determine the truth via consistency with axioms. However, the problem is that in this process, we have left out other modalities or interpretations of sentences which may or may not be true, and that leads to the incompleteness.

Therefore, once we understand the universalist language and how its problems of incompleteness can be overcome, we can extend that understanding to decipher the personal and culture specific linguistic nuances. I will not focus on the natural language problems in this book and limit myself to mathematical issues. These are the minimal requisite ground on which we need to build upon.

Book Overview

The main idea in this book is that Gödel's, Turing's, and other mathematical paradoxes are based upon two fundamental mistakes. I will call them *categorical* mistakes and *semantic* mistakes. The semantic mistake is that it is possible to call a symbol with meaning not-P by the name P. Similarly, it is possible to call a program not-P by the name P. When such names are interpreted as meanings, a contradiction is invariably created. The categorical mistake is that we don't distinguish between concept and name or between program and name. The paradoxes could be avoided if we treated not-P as a meaning and P as a name, and did not equate the name with the meaning. Contradictions result from a combination of categorical and semantic mistakes and can be avoided if either one of these mistakes is avoided. However, any attempt to avoid them in this manner leads us to a new problem—namely how do we distinguish between the different modes in which a word is being used. If we universally designate a specific mode of interpreting a specific word, we will miss several interpretations of sentences that employ other modes, and hence mathematics will be

incomplete. If instead we use multiple modes to operate simultaneously, we will create contradictions. This leads us to the familiar paradox of Gödel's incompleteness.

The problems of incompleteness entail that we need a view in
which both concept and name can be derived from the same world.
However, this raises some important questions about the *origin* of
the distinction between symbols. Are symbols originally just physical
things to which meanings are later added on? Or, are symbols primarily meanings which are given a physical embodiment later on? I will
show that if we begin with the idea that symbols are physical things
which are given meanings later, then we still have unanswered problems about how these things are *distinguished*. To distinguish things
we need some *types* in terms of which they can be distinguished.
For instance, two balls can be distinguished by their color or shape.
Thus, even the physical distinction between things requires a prior
existence of types. If, however, we assume a type distinction *a priori*
then the physical distinction follows. Therefore, the type distinction
is stronger than the thing distinction and we must acknowledge that
symbols are originally meanings even before they are given a physical
embodiment. The way they are physically distinct is hence how they
were already different types.

Beginning with an analysis of the incompleteness in Gödel's theorem, the book explores the consequences of solving the incompleteness in the foundations of mathematics. The second chapter deals with
Gödel's incompleteness and why it represents a categorical mistake. I
will sketch Gödel's proof and how it uses numbers as names, concepts
and things. The three uses of numbers are not in themselves wrong but
using them interchangeably—without overcoming the semantic mistake—is wrong. I will show how incompleteness is solved if we distinguish between names, meanings and things. This approach to Gödel's
theorem is similarly applied to Turing's paradox which involves a category mistake between data and a program. I will illustrate that when
this mistake is corrected, the proof of Turing's paradox does not exist.
While a complete solution to the paradoxes requires a fix for both
semantic and category mistakes, this solution is not described until
later. The second chapter only discusses how by avoiding the category
confusion itself we can overcome the paradoxical situations.

The third chapter extends the category mistake discussion to other paradoxes in logic, mathematics and ordinary language. This discussion is primarily meant to show that Gödel's and Turing's theorems have a broad philosophical undercurrent with implications for many areas even outside mathematics. Of specific importance is how the dissolution of category mistakes is indicated by problems in atomic theory which require us to view atomic objects as symbols of meaning rather than classical particles. This chapter analyzes the similarities between a semantic view and the mathematical description of atomic objects in current quantum theory.

The fourth chapter discusses numbers, and how they should be treated in new ways. If nature consists of symbols, then distinctions and the counting of these symbols entail a different view of numbers. This chapter applies semantic ideas to analyze and interpret different types of numbers—negative numbers, complex numbers, rational numbers, irrational numbers and natural numbers—as properties of symbolic objects, sets and propositions. The chapter explores whether numbers are quantities and argues that the quantity view leads to interpretive problems for many classes of numbers. The idea that numbers represent names and meanings, however, avoids these difficulties. These ideas are then used to propose a semantic view of space and time, where these are described hierarchically rather than linearly. In a semantic space and time, a number denotes a different type, not just a physically different entity. I call this semantic space-time view a Type Number Theory (TNT) and it can be used to build a foundation for mathematics free of various paradoxes.

The fifth chapter discusses the idea that the foundation of mathematics is not in the notion of objects but in the notion of distinctions which, when applied to space-time, create semantic objects. This leads to the idea that counting depends on a theory of types, and objects have to be derived from the types in the theory. The world of things has to be seen as a world of types. Distinctions between these types help us distinguish things. Once things have been distinguished, they can be ordered. It is this ordering that leads to counting, which then eventually produces numbers.

This book is about the role of semantics in mathematics and computation and how it relates to problems in mathematical foundations.

Mathematicians have ignored questions about meaning in mathematics. This book argues that ignoring the role that meaning plays in mathematics leads to problems of incompleteness and incomputability. The question of semantics is however not limited to mathematics. It is also important to understand issues about semantics from logic, philosophy and ordinary language before we can apply these ideas to mathematics. The solutions to these problems also require us to understand the nature of language and how our minds process and understand the world through language. The human mind does not encounter semantic issues, although the issues appear repeatedly in mathematics and in mathematical descriptions of nature. By having mathematics include semantics, and building physical theories that can compute meaning, every other domain of science can address issues related to semantics.

2

Gödel's Mistrick

If people do not believe that mathematics is simple, it is only because they do not realize how complicated life is.
—John von Neumann

A Barbaric Logic

Before I go into describing the problems in Gödel's proof, I will analyze the Barber's Paradox, quoted first by Bertrand Russell but attributed by him to another unnamed source[7]. This analysis of the Barber's Paradox will bring out some of the salient issues in Gödel's proof since the basic issues from the Barber's Paradox are also inherent in Gödel's proof. Both arise due to confusion between two aspects of language that must always be distinct but are easily confused in certain circumstances. The analysis of the Barber's Paradox therefore serves to identify the nature of the problem in Gödel's proofs. The Barber's Paradox is the following innocuous statement about a professional (and presumably male) barber.

A BARBER SHAVES ALL THOSE WHO DON'T SHAVE THEMSELVES.

This describes a commonplace scenario in which people visit a barber to get themselves shaved. If they had shaved on their own they would not need the services of a barber. Hence, a barber shaves only those who do not shave themselves. The paradox arises when we ask: Does the barber shave himself? And the commonsense response is that (if the barber is clean-shaven and male) that either the barber

43

shaves himself or some other barber shaves him. Yet, this common conclusion cannot be derived from the above statement in any simple manner. In fact, upon analysis, we find inherent contradictions in the statement, leading to the paradox. There are two possible alternatives for a real barber—either a barber shaves himself or he does not—and surprisingly both lead to a contradiction.

To illustrate, let's suppose that the barber shaves himself. Since the barber is supposed to shave only those who do not shave themselves, the barber should not shave himself. Conversely, if we suppose that the barber does not shave himself, then he must shave himself because the barber must shave everyone who doesn't shave himself. Either way, we end up with a logical contradiction.

In the current literature about the Barber's Paradox, the logical contradiction is seen as a consequence of a barber self-referencing himself in a statement involving a negation. In other words, we treat this to be a problem about how we employ words within language and nothing more. The problem would not arise, for instance, if the barber was a woman (then she does not need to shave herself), or if the barber is not clean shaven. The problem arises when the person shaving is the person being shaved, and this leads to contradiction, because we have the same individual as the subject and the object.

But self-referencing is not the root cause of the problem. Rather, self-reference arises from using the word 'barber' in two ways— as a *class* and as a *person*. The class 'barber' denotes people that "shave those who don't shave themselves" and it basically explains the meaning of the class 'barber'. There is, however, also a person barber, who belongs to the class 'barber'. This person acts as a barber most of the time, but on some occasions he might not. The paradox arises when we conflate the *class* barber with the *person* barber and ask—does the *person* barber act like the *class* barber? If you look closely, the times when a barber shaves himself are instances in which the person does not belong to the class. This same person might have earlier belonged to the class barber when he shaved others who do no shave themselves. If you forego the difference in time between someone being and not being a barber (by alternately shaving others and shaving himself) it appears that a person is both in the class and not in the class, which looks like a paradox. In reality, however, there is no such paradox, because a person can shave himself and then shave others. While shaving himself, a person is not a barber. Such a person can act as a barber when he shaves others. Even a skilled professional barber when shaved by another barber friend is not a barber at the point of being shaved. In both cases—whether the person shaves himself or is being shaved by another professional barber—there is no contradiction because he is not a barber at the point of being shaved. He may be a barber in some situations, which is entirely consistent with not being a barber in other situations.

The Barber's Paradox arises due to a basic issue in rational knowledge—namely that rational truths are expected to be timeless. If we apply the noun 'barber' to someone, sometimes, this knowledge must be timeless—"once a barber, always a barber." We cannot take it off for sometimes and then apply it back again. To say that "X is Y" means that X is always Y, at all times, all places and all circumstances. The paradox arises when we apply the 'barber' attribute to a person even when he is not acting like a barber, although he may have acted so in the past. Therefore, a person who shaves other people during the day but shaves himself in the morning would be a barber during daytime, but not in the morning. The common noun 'barber' can be applied to this person when he is shaving others and not otherwise. Because the person is not always a barber, he could shave himself without creating a logical contradiction. The contradiction arises only when we equate the *person* with the *class*, timelessly.

The word *'barber'* can denote both an individual person and the class of individuals and this is the source of the language twist. The word 'barber' when interpreted as a class denotes a concept. The same word when interpreted as a person represents an individual. The individual and the concept are correctly related when the individual is acting according to the concept, but this might not always be the case. A person bears out the concept when he shaves other people, but not when he shaves himself. At the point of shaving himself, he is by definition, not a barber. The contradiction can be solved by distinguishing the *person* from the *class*, or the conceptual from the physical, which are two ways to use the word 'barber'.

Gödel's Numbering

Like the Barber's Paradox that arises through a confusion between two kinds of uses of a single word, Gödel arrived at the incompleteness proof by using the same number in two ways—as concepts and as names. Like the word 'barber' has two kinds of interpretations, numbers too can denote two kinds of things. Numbers can be used to *name* objects. For example, we might call someone 'Solider 343634' or 'Employee 98723'. These numbers refer to a specific individual. Similarly, we might call some statement as 'Statement 578930', which names a specific statement. Gödel created a scheme where he labeled statements by numbers, like how we would number soldiers or employees. In this situation, the number *refers* to a statement through a name. But since the statement also has a meaning, we could (erroneously) treat the name that refers to the statement as also indicating its meaning. '578930' now represents the *meaning* of a statement as well as a *name* to individuate it. This is analogous to how the word 'barber' can name a person as well as his type.

The scheme to label statements with numbers produces an effect like the Barber's Paradox. Quite like the term 'barber' can name a person or denote a concept, a number now represents two things. Ideally, we should distinguish between these two interpretations of a number, but this distinction doesn't exist in mathematics. Gödel created names for statements and interpreted them as meanings. This is like how we might call someone Mr. Barber and then expect him to shave others due to that name. Not everyone who is called Mr. Barber may be a barber, and this leads to a paradox when Mr. Barber shaves himself. The ability to interpret a number as a *meaning* and as a *name* underlies the paradoxes that led to incompleteness. But this is an error in how numbers are used. Just as a Mr. Barber is not necessarily a barber, similarly, every statement called by a number may not exhibit that number. The name is a convenient label to refer to that statement, and because we choose to name a statement by number does not give it the properties of that number.

A name is the outcome of a method of physically distinguishing statements. It depends on the language (English, French, German) used to express the statement and the specific words employed in it. If the vocabulary underlying the statement is changed, then their physical mappings to numbers will change as well. The same meaning would then be called by a different name. A numeric label of a statement is not identical to its meaning and to use that label to understand the meaning is only context-sensitively true. The distinction between names and objects is given in the saying 'A rose by another name would smell as sweet'. Gödel in effect named a rose as a marigold and then expected it to smell differently. If we expect the change in name to change its meaning, a contradiction arises.

The mapping between names and meanings can be changed in two ways. First, we can think of a name-meaning mapping change as a coordinate transform that alters the labels associated with objects. Since a coordinate transform does not change the *relative* positions or relations between objects (or their physical properties), which determine the meaning, by changing the coordinate system, we have changed the *mapping* between name and object, without changing the object's meaning itself. Now we can describe the symbols by a new name without changing the symbols themselves. For instance, we can perform a coordinate transform such that a symbol not-P is mapped to a number P. Second, we can fix the coordinate system and move a symbol within a coordinate system such that it is now located at a new location that contradicts its meaning. For instance, we can move a symbol not-P such that it is located at point P.

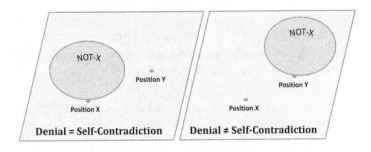

Figure-3 The Problems of Physical Numbering

Gödel, however, did not use either of these methods. Gödel fixed the coordinate system and did not move statements around. He was still able to find statements not-P that map to a number P. This is because in a physical space the distinction between the symbols is *not* based on their semantic difference but based on their physical attributes (the same meaning can be physically expressed through different words and the expression will be called by a different name or number although it has the same meaning). The only way to fix this problem is if we can fix the physical distinction between symbols based on their semantic distinction. Thus, the symbol with meaning not-P must always be at location not-P. If we move symbols, it should not just change their physical distinction but also change their meaning. Thus, it would be impossible to move the object "Not-X" to a position X. In trying to move the object to a new location, we will change its meaning, making the contradiction impossible.

We have briefly discussed this scheme earlier, and I will elaborate on it further in the subsequent chapters. We need to think of the world as symbols distinguished by their meanings. This semantic individuation between symbols is *reflected* in the coordinate system as names, and the meaning-to-name mapping is fixed. This addresses each of the above three problems:

- The symbols cannot be moved around arbitrarily because the location of a symbol depends on its meanings. This name-meaning mapping represents an 'absolute' coordinate frame.

- We cannot arbitrarily change the coordinate system, without changing the name-to-meaning mapping as given by the absolute coordinate system defined above.

- We can never find a symbol whose physical distinctness contradicts its semantic distinction.

For now, let's return to how Gödel maps meanings to names using a physical (rather than a semantic) numbering system. Gödel numbering maps a symbol, statement or formula to a natural number, which is called its Gödel number. To create numerical names for sentences, Gödel took two steps. First, he assigned a unique natural number to each basic symbol in language. Second, he devised a scheme to encode an entire formula (a sequence of symbols or a statement) into a natural number. This can be understood through an example. For instance, we might map the alphabets in English onto numbers from 1 to 26 and the word 'barber' would then give us a sequence of digits as {2, 1, 18, 2; 5, 18}. This sequence can in turn be converted into a single number using other schemes that map the number sequence to a single number. Gödel also employed a mapping scheme that converts a statement into a unique number. His method of generating a unique number is shown below.

$$GN(x_1 x_2 x_3 ... x_n) = 2^{x_1} 3^{x_2} 5^{x_3} ... p_n^{x_n}$$

Here, $x1, x2, x3$ represent individual symbols and their sequence represents a mathematical formula, conjecture, axiom or theorem. Using this method and given a mapping between letters and digits, we would represent the word 'barber' as $2^2 * 3^1 * 5^{18} * 7^2 * 11^5 * 13^{18}$, which is a number with nearly forty digits in it. According to the Fundamental Theorem of Arithmetic[8], any natural number can be expressed as a product of primes and Gödel used this property in numbers to create a unique mapping. Note that in the numbering scheme, a prime number denotes the position of a letter in a statement and the letter at that position contributes to the exponent of the prime number. Any letter will therefore contribute a different value based on its position in the statement. The scheme produces unique numbers from statements[9]. Given this uniqueness, we can say that a Gödel number for a statement *names* that statement because each name will uniquely identify only one statement. By converting a string of letters to natural numbers we facilitate their manipulation in arithmetic. However, these numbers are names for statements and they do not give us the meaning of the sentence. At best, the number tells us about the syntactical representation of that meaning in a sentence. But in mathematics the syntactical representation also defines the meaning because there is no other way to determine a statement's meanings. Thus, the numbering scheme was used by Gödel to represent both a name and a meaning. When the name does not correctly identify the meaning, paradoxes can occur.

Gödel's numbering depends on the alphabets being used in a statement and how alphabets are mapped to numbers. The numbering does not depend on meanings in a sentence. If we change alphabets, the number of alphabets, or the alphabet to number mapping scheme, Gödel numbers will also change. This means that the statement "A barber shaves all those who don't shave themselves" in French will have a different Gödel number. The Gödel number represents how a statement is *physically* expressed, although the number does not represent the statement's meaning. The meaning of a statement is independent of whether the statement is written in English or French, although its representation depends upon the language. Gödel's number measures the properties of a statement's linguistic expression. Ordinary language statements or mathematical theorems are however semantic objects although they need a language to express them physically. Different languages will create different representations of the same idea. Gödel's numbering maps propositions to unique names, leaving the meaning untouched. But when a number used in a representation is interpreted semantically, it seems that we are talking about meanings although the number was originally only a physical representation.

The fact that Gödel's number pertains to the physical expression of meaning (and not to the meaning itself) is the primary idea we need to understand to comprehend the fallacy of Gödel's argument. The use of a specific formula that Gödel devised to map statements to numbers is secondary to his argument, so long as it yields a unique mapping. The numbering scheme could as well use a different formula. Yet, each such scheme will measure the physical embodiment of a meaning or its physical expression in language.

The key points of the subsequent discussion therefore don't depend on the exact scheme of numbering that Gödel devised, but only on the fact that Gödel used a numbering scheme to uniquely map statements to numbers. The sophistication in Gödel's scheme is designed to keep the mapping between statements and natural numbers unique. In later sections, therefore, I will use a much simpler scheme to illustrate the basic structure of Gödel's argument. My scheme will not bother with the uniqueness in the number-statement mapping as this fact is irrelevant to the main issue. My scheme will basically just map a sentence to the number of letters in it. In effect, the scheme will measure a statement's length as a physical property and map sentences to these lengths. The scheme is not unique because all identical length sentences would be mapped to the same number. But the scheme captures the salient points of Gödel's argument and makes it easier to understand the crux of the issue at hand, without a complicated computation of numbers.

Deconstructing Gödel's Proof

Gödel's proof hinges on the construction of contradictory statements. A common example of such a statement is 'This statement is false'. This can be freely created and is meaningful, so long as the word 'this' in the statement refers to another statement besides itself. However, a problem arises when 'this' refers to the statement itself, because in that case, the statement becomes logically self-contradictory. Gödel's proof constructs a statement 'This statement is false' and then makes 'this' point to itself. The self-reference is achieved by equating meaning with naming, as follows. Note that 'this' is a *name* by which we can refer to a sentence when the word is used to point to a sentence different from itself. The referred statement here can be 'The statement is true'. So, we have a statement which claims that it is true, and another one that claims that it is false. One of the two statements ought to be false, if the logical system is consistent.

This statement is false

↓

The statement is true.

Here, 'this' in the first statement points to the second statement. The meaning of the word 'this' in the first statement is the second statement. The second statement claims to be true. The second statement thus has a *meaning*, namely that it is true. Separate from these meanings is the truth condition of the two statements. For instance, the second statement could be false, and that would make the two statements consistent. Alternately, the first statement could be false, which will also make the two statements consistent. If we have two statements, and we can distinguish between their meanings and truths, we don't have a real problem. The problem however is created if we substitute 'this' in the first statement with the second statement creating the following complex statement.

'The statement is true' is false.

This statement has a single meaning and a single truth condition. Its meaning is that the subject of the statement is false. But we cannot be certain if this statement is true or false. If we say that the statement is true, then its meaning is that it is false. If we say that the statement is false, then its meaning says that it is true. We now have a conflict between the meaning and the truth of the statement. If the meaning is true, then the truth is false. If the meaning is false, then the truth is true. Note how this problem arises when we convert two statements into one, by substituting 'this' with a statement.

The substitution of 'this' with a statement is non-trivial because 'this' is a name that refers to an object (the sentence). By substituting 'this' with its referred object, we have committed a categorical mistake of replacing a name with its referenced object. The meaning of the name is that it points to a statement, but the meaning of the statement is that it claims to be true (the statement could have other meanings). These are different classes of meanings and cannot be substituted. They are as different as Mr. Barber which refers to an individual and 'barber' which refers to a conceptual class of things.

In effect, the problem amounts to saying that if Mr. Barber does not shave others then we have contradicted the meaning of barber. The plain fact is that one doesn't follow from the other. The Barber's Paradox too can be deconstructed into statements: (a) Mr. Barber does not shave others, and (b) A barber shaves others. So far, there is no logical or conceptual problem. The problem arises if we throw the equation 'Mr. Barber is a barber' into the fray. Now, we end up with the statement (after substituting Mr. Barber with barber in the first statement) 'A barber does not shave others' which contradicts the second statement 'A barber shaves others'. In the above example, I have distinguished a name from a meaning by calling them Mr. Barber and barber, but that is only for our ease of understanding. We can easily see how the same word (barber) can denote both a conceptual meaning and a specific individual which underscores the fact that the same symbol can be used to represent different kinds of entities. Unlike ordinary language, the ability to distinguish between different meanings of a symbol doesn't exist within mathematics. It just means that we have made it easy to create logical paradoxes.

Gödel picked a statement that said 'P is false' and then labeled this statement P. Let's distinguish these two as a name P_N and as a concept P_C, respectively. P_N is a proper noun since it calls out a statement by a name. P_C on the other hand is a common noun as it describes the properties of the statement as its meaning. But both common and proper nouns can be represented by the same symbol. This is as true in mathematics as in ordinary language. And yet, in ordinary language we carefully distinguish between common and proper nouns even when the same word is used to express them. We are not able to do so in mathematics because the field does not directly deal with meanings. Gödel effectively abused this freedom to represent both common and proper nouns by the same symbol. His trick amounts to picking a random statement that says "this person does not shave others" and then labeling it as Mr. Barber. There is no self-contradiction unless we equate Mr. Barber with barber. In Gödel's case the contradiction arises if we equate P_C with P_N. That is, the contradiction arises if we conclude that when P_N calls a statement, it is calling itself. This self-reference and consequent self-contradiction are an outcome of equating meanings with names, a fallacy like thinking that every Mr. Barber is also a barber.

This fallacy is easily committed in mathematics because 2 can denote the meaning of twoness and the name that refers to the 2nd object. The 2nd object may not exhibit the property of twoness, and we would be committing a mistake if we expected this equivalence. To illustrate, assume that we are ordering a set of objects that have even numbered masses, beginning with 2 kilograms. The first object is 2kg, the second object is 4kg, the third one is 6kg, etc. Now, the second object doesn't have the property of twoness because its mass is 4kg. This does not create a self-contradiction unless we equate the ordinal number 2 with the cardinal number 2 and expect the second object in the order to have the property of twoness. Congruence between ordinal and cardinal numbers exists in natural numbers but may not exist in general. That, however, does not entail a logical contradiction, unless we equate the ordinal and cardinal.

If a barber is called Mr. Barber the name is consistent with his properties, so long as Mr. Barber does not shave himself. However, if a porter is called Mr. Barber then there is an inconsistency between the name and the behavior. Of course, in ordinary language and the everyday world we have no problem in calling a porter by the name Mr. Barber because we distinguish between the common noun and the proper noun. The common noun defines generic properties and the proper noun defines a specific object. Calling a porter by the name Mr. Barber may be a little confusing or inappropriate but it is not logically inconsistent. But in mathematics we can't distinguish between common and proper nouns. Without the distinction between meanings and names, we are likely to make mistakes.

The earliest example of such a mistake was given by the Greek philosopher Epimenides who said "All Cretans are liars" although he was a Cretan himself. The statement represents a logical paradox because if we believe what Epimenides says to be true then his assertion entails that he is a liar himself. Since Epimenides was a Cretan, and by his own admission all Cretans are liars, then Epimenides must be a liar, too. If, however, we say that Epimenides is a liar, then what he said about Cretans must be false and therefore all Cretans must be truthful. If therefore Epimenides is truthful then he is a liar. If, instead, he is a liar then he is truthful. In the Epimenides paradox the word 'Cretan' plays a dual role. It acts as a *name*—pointing to Epimenides—and as a *meaning*, indicating all people living in Crete. We create a self-reference by collapsing the distinction between the name and the meaning. This is a category mistake. If, for instance, Epimenides was not a Cretan when he made that statement, or if he was born outside Crete but lived in Crete when he made that statement, the contradiction would not exist.

Logic and Category Mistakes

In logic, if A is B and A is C then we equate B with C. For instance, if A is B ≡ 'All Cretans are liars' and A is C ≡ 'Epimenides is a Cretan' then by the logic of equals, we would conclude that B is C ≡ 'Epimenides is a liar'. This conclusion rests upon equating the name Cretan that calls out an individual Epimenides with the type Cretan which is a class referencing the people living in Crete. This equation works in many cases, but not in all. For instance, the equation would fail if we invert the relation between Cretans and liars—that 'All Cretans are liars' does not imply 'All liars are Cretans'. In 'All Cretans are liars' liar is a class and Cretan is a member of the class. In 'All liars are Cretans' the relation between class and object is inverted and Cretan is a class and liars are members of that class. The class of Cretans has many properties, of which lying may be one. Thus, equating a class with one member leads to an error. We can say that an individual in a class underdetermines the class, if the class has other members.

If instead there are many classes that apply to an object, those classes underdetermine the object. For instance, if A is B ≡ 'John is a doctor' and A is C ≡ 'John jogs every day in the morning', it would be erroneous to conclude that B is C ≡ 'Doctors jog in the morning'. Jogging and doctor are common attributes of John (an individual) but they are not semantically related and can't be logically equated. Similarly, if A is B ≡ 'Aristotle is Plato's student' and B is C ≡ 'Aristotle is Alexander's teacher' the syllogism leads us to the conclusion that 'Plato's student is Alexander's teacher' which is false because Plato had many students who were not Alexander's teachers.

In ordinary language we frequently apply concepts to objects. While this application helps us draw useful conclusions in many cases, dropping the distinction between concepts and objects can lead us to erroneous conclusions. This is because a concept may apply to many objects and an object may be described by many concepts. Unless there is a single object that completely represents the meaning of a concept and there is a single concept that completely describes an object, objects and concepts cannot be substituted.

The distinction between concepts and objects is not always easy if they are represented by the same word or sign. For instance, the word Ford may represent a class of cars manufactured by the Ford Motor Company. The word may also name a person Henry Ford. While a car may be Ford because it is manufactured by the same company, the person Ford is not a car. Without the distinction between the two uses of the word Ford, we can be led into the false conclusion that if a car is Ford then Ford must be a car.

The distinction between different uses of a number doesn't exist in mathematics and this lies at the root of Gödel's proof, which uses numbers in two ways: as a name to identify a statement and as a meaning of that statement. Gödel's numbers have a dual use in his proof: they denote both a sentence's meaning and its physical embodiment within the sentence. As a name, the number calls out or refers to a specific sentence. And, as a meaning, the same number represents the concept embodied in the sentence. In mathematics we cannot distinguish between the specific sentence and the meaning it conveys. Without this distinction, we can create paradoxes if we blindly substitute the meaning with its physical representation.

Blind substitutions of names and concepts constitute category mistakes, but they are not unique to mathematics. While we saw how they lead to erroneous conclusions, the same trick can also be used to create logical contradictions. Consider the statement 'this is not red' where 'this' points to a red apple. If we substitute 'this' with the red apple, we are led to the self-contradiction that 'red apple is not red'. As separate statements, a statement that refers to a red object and claims it is not red is false, although there is no self-contradiction. By replacing a name in a false statement with the referred object whose facts contradict the claims in the statement, we create a self-contradiction. Self-contradictory statements are undecidable. Gödel committed this category mistake when he equated a name that refers to a statement with the statement's meaning. The name and meaning can appear in separate sentences and they will denote different things without a logical contradiction. By collapsing two statements that are mutually contradictory into one, we create a self-contradiction. This is not a property unique to numbers but can be easily achieved within any type of language whatsoever.

What Is Completeness?

There are two ways in which we can construct statements. First, a statement can be constructed by freely combining words to form sentences such that these conform to the rules of grammar. Learners of a new language learn meanings by using words in sentences. Freely constructed sentences aren't always true, but they are generally meaningful, provided they conform to the rules of language. Second, a statement can be derived as a conclusion using logical deduction from premises. If the premises are true, then the conclusions are also automatically true. False statements can be derived from true ones using negation. Completeness is the equivalence of these two methods. That is, every statement that can be formed through grammar can also be obtained via logical deduction, either directly or through a logical negation.

In ordinary language, these two methods of forming statements are *not* equivalent. For instance, "the sky is purple" is a grammatically correct and meaningful statement but it is not true. We can deny this statement through empirical evidence. In a formal system, such as mathematics, the same thing could be done using logical deduction. Assuming that mathematics is complete, and the universe can be described mathematically, it would be possible to refute this claim *logically*. That is, it should be possible to derive the statement, "the sky is *not* purple," from some suitable axioms about the world, which would logically contradict the claim constructed using vocabulary and grammar. Contradictions would automatically demonstrate that the meaningful claim is logically false.

Hilbert's program was to demonstrate that all freely constructed statements could also be logically proved or disproved, if the axiom system was complete. This follows from the completeness of mathematics. The only seeming hurdle to doing this would be the inability to form a complete set of axioms, and Hilbert assumed that wasn't the case. The completeness of mathematics was therefore defined as the equivalence between the sets of grammatically correct statements and logically provable and disprovable statements.

This point is crucial because, in the proof for mathematical incompleteness, Gödel used freely constructed, grammatically correct statements, which were not derived using logical deduction. Since these statements conform to the rules of grammar, they are meaningful. Gödel then showed that some such freely created statements can be logically contradictory, and these statements cannot be proved or disproved. Gödel interpreted the inability to prove any freely created statement as the fact that the axiom system is incomplete because completeness is defined as the ability to prove or disprove through logical deduction any grammatically correct statement. Gödel showed that every statement that conforms to the rules of grammar need not be logically provable. For instance, the statement 'I am false' is grammatically correct but cannot be proved. Gödel's proof can be interpreted as an indication that grammatical construction and logical proof are not equivalent. That is, it is possible to make meaningful statements that can't be proved.

Every language comprises some alphabets and grammatical rules by which these alphabets are combined into sentences. By complying with rules of grammar we create *meaningful* statements[10]. Once we have several meaningful statements, we could designate some of them as *axioms*—i.e. *a priori* true. Hereafter, axioms can enter logical inferences, telling us whether any randomly picked statement is consistent with the axioms and hence true. There are now two ways to create sentences: (a) by using word dictionaries and rules of grammar, and (b) by using logic to derive them from basic *a priori* axioms. Grammatically correct statements must have a subject and an object—they may include verbs and adjectives. A logical inference uses rules like 'if A is B and B is C then A is C'. Grammar rules make statements meaningful but there can be meaningful statements that are not compatible with axioms. Whether or not a statement complies with axioms is given by the rules of logic.

There have been attempts by mathematicians to dissolve the distinction between logic and grammar as part of symbolic or mathematical logic. The idea is that both grammar and logic deal with what is possible. A language is comprised of alphabets which appear in mathematics as axioms. How we create grammatically correct sentences must be equivalent to how we derive sentences using logic. This is largely a fair assumption, except that there is one thing that is available in grammar although not in logic. This is the ability in grammar to produce statements *about* other worldly facts or even about other statements. Logic only produces statements without any reference or aboutness[11]. In short, grammar permits and uses the subject-object distinction, which does not exist in logic.

Logical statements must be verified against axioms and their truth is ensured by logical construction which determines their meaning. However, it is possible to separate meaning and truth in grammatically correct statements that make claims *about* other statements or about other objects. Meaning in a grammatically correct statement is different from the truth of the statement it refers to; the meaning is in the referring statement and the truth depends upon the referred statement. Gödel used sentences that were grammatically correct. He picked statements about other statements such that the referring statement *denies* something about the referred statement. Although the referring and referred statements were different, Gödel mapped both referring and referred statements to the same number. The referring statement refers to the referred statement by a name, which was the name for itself. The statement now appears to deny itself, causing a self-contradiction.

But, in another sense, Gödel's definition of completeness is also flawed. He defines completeness as the equivalence between grammatical and logical constructions, but they aren't equivalent because grammar allows aboutness whereas (first-order) logic doesn't. This distinction would have gone unnoticed if mathematics used the distinction between meaning and naming and thus using logic as in everyday language. But Gödel assumed that grammar and logic are equivalent, thereby introducing aboutness in mathematics and then violated the distinction between meaning and naming that follows aboutness. Statements about statements can't be proved or denied logically; they must be proved or denied empirically. In a language that is dealing with aboutness there must be a difference between how a symbol is used as a reference to an object and as a meaning. The trick lies in bringing aboutness into mathematics when this is not possible in logic and then not honoring the distinction between meaning and objects that comes with aboutness.

Gödel defines completeness as: "For a formal system S in formal language L, S is semantically complete or simply complete, if and only if every logically valid formula of L is a theorem of S". This definition is equivalent to the following statement in ordinary language—"If a sentence can be constructed using the English vocabulary, such that it conforms to English grammar and word conventions, then that sentence is either *logically* true or false." Here, a formula in mathematics is a sentence in English, the language is English vocabulary and the formal system is the collection of all possible English sentences. A theorem must be proven by logical deduction, but a "valid formula" need not be constructed through logical reasoning; we can randomly create it by employing rules of grammar. A referential statement cannot be logically proven, although it can be constructed using grammatical rules. Logical proofs are not enough for referential statements because the truth of referential statements depends on the reference and not just the axioms. However, when a statement refers to itself, the referential truth is converted into a self-reference, allowing for a logical treatment of the statement's origin. This conversion, however, commits a category mistake.

The Structure of Gödel's Proof

Since a complete system must be able to prove a randomly formed statement, Gödel begins by formulating a random statement:

P CANNOT BE PROVEN WITHIN AXIOM SYSTEM S

This is a valid, free construction, as a statement about another statement *P*, although we can't tell at this time whether the statement is true or false. We also don't know anything about the statement *P* that is referred here. Gödel now maps the above statement (using Gödel's numbering) to a natural number. Let's assume that Gödel's number for the above statement is the number Q. The above statement about a statement *P* can now be written as follows:

Q: P CANNOT BE PROVEN WITHIN AXIOM SYSTEM S

In this form, statement *Q* makes an assertion about a statement *P* and we still don't know if *Q* is true because we don't know what *P* is. Now Gödel applies a trick. He tweaks numbers *P* and *Q*, until they become identical. It involves selecting a number in a statement such that the Gödel number for the whole statement is the same as the number used within the statement. The result now, is a self-contradiction.

N: N CANNOT BE PROVEN WITHIN AXIOM SYSTEM S

If we presume that the above statement is false, then it is provable. If instead it is true, then the statement cannot be proved. The first position implies that the axiomatic system is inconsistent since false statements can be proved. The second position implies that the axiomatic system is incomplete since true statements cannot be proved. Like I said in the opening portion of the book, mathematicians generally prefer incompleteness to inconsistency: it is better to have true statements that are unprovable instead of having false statements that can be proved. Thus, Gödel's theorem is said to indicate mathematical incompleteness, and not inconsistency.

If we look at the argument closely, we will see that the number N is used in three different ways. In the statement "N cannot be proven within Axiomatic System S," N represents a *token*. The statement represents a meaning *not-N* or *N-is-false*. As "N:" N represents a *name*. The numbering scheme labels a sentence with a number which is a name for the sentence. Thus, when we instigate a self-contradiction, we are confusing different uses of the symbol N—a name, a meaning and a token. The problem is like using the same symbol 'barber' in two different ways—one as a *name* and the other as a *person*. However, unlike the barber case, where we intuitively understand the nature of barbers, we cannot do the same with N—since we don't know what N means and does.

The key problem associated with Gödel's theorem is sometimes perceived to be a self-reference, where a number outside the statement refers to a number inside the statement. In this view, the ability to create self-references means that we can never unwind the meaning of a statement as the meaning is in turn to be expressed by expanding the statement and this leads to an infinite recursion. Since N in the statement "N cannot be proven within Axiomatic System S" is a symbol referencing another statement, to fully elaborate the statement, we should substitute the symbol with the actual reference, which leads to a statement "Statement 'Statement N cannot be proven within the Axiomatic System S' cannot be proven within Axiomatic System S". But we still have not resolved all references in the statement and need to substitute N with its description. Trying to expand a reference leads to an infinite regress, since every description of *N* must also in turn refer to *N* and a mechanical procedure can never prove or disprove such a statement.

A Simplified Argument

Gödel's argument can be simplified to illustrate the confusion between name and meaning. In the below simplified form of the argument, all details about Gödel's numbering are irrelevant. I have replaced them by a simpler numbering scheme that maps statements to the number of letters in them. Consider the statements below:

Statement X: This statement has 33 characters.
↓
Statement Y: *Your dog barks non-stop at night.*

Let's suppose that we map statements to numbers based upon how many letters they have. Both above statements contain 33 letters each and they would be mapped to the same number (33).

33: *This statement has 33 characters.*
↓
33: *Your dog barks non-stop at night.*

Now, if we remove the second statement, then 'this' in the first statement points to itself. And there is no problem because statement X indeed has 33 characters and it can point to Y or to itself without logical inconsistencies. The statement now looks as follows.

33: *This statement has 33 characters.*
To illustrate Gödel's trick, now consider the below two statements:

33: *This statement has 35 characters.*
↓
33: *Your dog barks non-stop at night.*

Both statements still have 33 characters. But the first statement is false which shows the difference between physics and semantics. Physically, the statement has 33 characters, but the statement claims that another statement has 35 characters. This is a false claim, but not logically contradictory. If we remove the second statement, then 'this' in the first statement appears to point to itself. Now there is a contradiction between the physical composition of the statement and its meaning because the statement has only 33 characters although it claims to have 35 characters. The statement now is:

33: *This statement has 35 characters.*

This is like Gödel's proof of incompleteness: "Statement P: P cannot be proven within Axiomatic System S". Gödel's scheme works by converting a false claim about another statement into a false claim about itself, making things appear self-contradictory. The numbering scheme that maps referred and referring statements to the same number converts a false claim into a self-contradiction. False claims are possible in mathematics, but self-contradictions are not. Gödel uses a false claim to arrive at a self-contradiction. The solution is to limit the statement to a false claim and not allow it to become self-contradictory. To do that, we must view numbers semantically such that 'Statement X' maps to X because they have the same meaning. In this scheme, 'Statement X' and 'Assertion X' will also map to the same number X, because both have the same meaning. In short, the mapping to numbers depends on meaning rather than letters.

Contradictions can exist in nature, but self-contradictions cannot. A self-contradiction implies that the statement physically exists, but its meaning denies its existence. Mathematicians have constructed such statements because the statement construction did not consider the statement's meaning. A statement is only a physical thing, whose meaning is only present in the mind. A mathematician can take a token, map it to a name, then convert that name into a meaning and apply it back to the token. The genesis of this problem is in naming, which cannot be prevented in current mathematics because there is no specification of how things must be named. In current mathematics, any object can be called by any name. To prevent contradictions, there must be semantic rules that limit the possible names that can be associated with a thing. For instance, a stone can be called a chair or table, but not a flower. Also, under no circumstance can we call something that is a stone by the name 'not-stone'. In other words, to prevent contradictions, we need semantic numbering procedures that map things to names in accordance with their physical properties and consistency between names and things. We have earlier discussed such a scheme and we will return to that discussion in greater detail in the later parts of this book.

Disproving Gödel?

Mathematics lends itself to two broad kinds of interpretations: (a) it is viewed as pure ideas by practicing mathematicians and (b) it is seen as physical tokens by computer scientists. Under the Platonic assumption that the present world reflects the pure world of ideas, the two appear to be interchangeable. But what if the present world needs both ideas and matter, such that the two can co-exist within a symbol? Now, the solution requires fixing the *category mistakes* and *semantic mistakes*. By fixing the semantic mistake, a concept not-P can never be mapped to a name P. By fixing category mistakes, we would distinguish between the name and the meaning. Ideally, we need to fix both category and semantic mistakes, not just one. But as we have seen earlier, if we fixed the semantic mistake, we could retain the category distinctions but permit category interchanges.

Completeness requires a new philosophical view in which the physical distinction between symbols is derived from their semantic distinction. A concept P will thus have a name P and a concept not-P can never have a name P. With this change, the categorical mistakes will become irrelevant and we could use mathematics like we use it today—i.e. without caring about the categorical differences. Can we now disprove Gödel and show that mathematics is complete? Refuting Gödel's proof does not amount to proving that mathematics is complete, as Hilbert envisioned, because completeness needs a positive confirmation not just negation of its impossibility. It can, however, be said that a language that avoids either category or semantic mistakes will not prove Gödel's incompleteness.

We might say that Hilbert's Second Problem is still open, although the problem is not limited to mathematics. Rather, we are asking questions about language itself: Can language consistently encode meanings? If ordinary language is complete, then formal languages that incorporate features of this ordinary language are also complete. Alternately, a complete formal language with features of ordinary language will show the completeness of ordinary language. Completeness is thus not a question unique to mathematics but rather a question of incorporating basic distinctions between naming and meaning, in any language. Alternatively, if such distinctions cannot be provided, at least the names must be identical to meanings. As we have noted, the problem that Gödel exposed exists in ordinary language as well. Indeed, the solution to the problem can be more easily seen in ordinary language than in mathematics.

Semantic Arithmetization

Gödel's numbering is a type of arithmetization by which a statement is converted into a natural number. However, Gödel's scheme performs arithmetization of *syntax* rather than of *semantics*. In Gödel's approach, the same meaning when expressed in a different language will have a different Gödel number. Similarly, the Gödel number for 'Sentence X' and 'Statement X' will be different although they have identical meaning. Imagine now a numbering scheme that converts statements to numbers based on their meanings rather than syntax. In that scheme, two statements with the same meaning but one in English and another in French will have the same Gödel number. 'Sentence X' and 'Statement X' will have the same Gödel number. Now, we cannot perform Gödel's diagonalization trick. That is, we cannot form the statement of the following kind:

STATEMENT N: STATEMENT N CANNOT BE PROVEN WITHIN AXIOM
SYSTEM S

The above statement will be illegal in semantic arithmetization because 'Statement N' and 'Statement N cannot be proven within Axiom System S' have different meanings and hence will map to different numbers. Since they map to different numbers, they cannot be equated, and the contradiction cannot exist. As it currently stands, Gödel's arithmetization converts statements to numbers based on the choice of symbols, the order of words and letters in a word, which are all syntactical elements of the statement. In this scheme, the arithmetization doesn't represent meaning but the physical properties of the statement, although Gödel treated the arithmetization as a representation of the statement's meaning. This led to the confusion between meaning and physical properties, which can be avoided either by distinguishing two uses of meaning, or by employing semantic instead of syntactical arithmetization.

Semantic arithmetization doesn't exist in mathematics today. Later in the book I will discuss a geometrical scheme where names and meanings are derived from the location of a symbol in space. This scheme requires us to treat the space of symbols *semantically*. That is, a different location in space represents both a physical and a semantic distinction. A different location in space is a different *type*. Using this scheme, it is possible to represent the meanings of objects as numbers, consistent with their syntax. The relation between a word and a meaning is invariable in this language, which means that the word-meaning association cannot be changed. This constitutes a universalist language suitable for mathematics where it is convenient to not complicate the subject with categorical distinctions.

Turing's Problem Revisited

The connection between mind, machines and mathematics needs more than a solution to Gödel's incompleteness. Incompleteness involves a descriptive use of language, when language represents knowledge about the world. Languages however aren't just used descriptively. A language can also be used to create programs which are *instructions*. The relation between the mind and intelligent machines depends on instructional rather than descriptive languages. The problems that

Gödel's theorems highlighted now appear in a different form. When language is used for descriptions, the key mathematical issue is to maintain a separation between a statement's meaning and its names. But when language is used for program instructions, the program's meaning is the actions done by an instruction and the name is its location in the program.

For instance, the solution to the 'Traveling Salesman Problem' will be called the 'Traveling Salesman Program' and the meaning of the program is the instructions that solve that problem. The key problem of computing languages is now as follows: Just as we would like to know if some meaningful statement is true, we also would want to know if a program will solve its intended problem.

Algorithms are procedures created to solve problems. Algorithms that solve a problem will provide a path from axioms to a solution, if the algorithm is correct. But how do we know that the algorithm is correct and will indeed take us to the intended solution and not somewhere else? How do we validate the correctness of a program like we validate the truth of a statement? How do we know that a program will achieve the purpose for which it has been designed? This problem can be divided into two parts. First, we need to know if the program will *halt*. Second, we need to know whether the program upon halting would have achieved its *goal*.

If a program achieves its goal, then we also know that it will halt. However, by knowing that a program will halt, we cannot be sure whether the program has achieved its purpose, although if the program halts, we can use another program to ensure that the first program's input matches with the intended output. The latter problem can be solved independently of the first problem. Therefore, Turing focused only on the question of whether the program will halt. Turing called this the Halting Problem and proved that there is no program that can determine if a program will or will not halt. It is one thing to know whether an algorithm solves the intended problem. It is another thing to know whether the algorithm trying to solve a problem (even an unintended one) will halt. Both problems are unsolvable; Turing's proof concerns only the latter problem.

Both problems are semantic. First, we have an *intentional gap* in that we don't know the specific problem that an algorithm solves. If we knew the specific problem an algorithm is supposed to address, then we could also know if it indeed addresses the intended problem. That would give us the ability to decide if we are solving the right (intended) problem through an algorithm. Second, Turing's Halting Problem is the issue that we don't know if the algorithm that is solving a problem is solving it correctly. If we knew that, then we would also know that it will come to a halt when it has solved it. So, we have two issues—we don't know if we are solving the right problem and we don't know if we are solving the problem rightly. Knowing that we are solving the right problem depends on the relation between a program and its intended goal. Knowing that we are solving the problem rightly depends on the internal correctness of the program itself. In computing terminology, the question of solving the right problem is called *validation* and the question of solving the problem rightly is called *verification*. Since automated procedures cannot verify nor validate, manual test efforts must be applied to confirm software for intents and internal correctness.

Algorithms that stop but don't solve the intended problem are meaningful statements that convey something different from what they claim to achieve. On the other hand, algorithms that never stop are syntactically correct although meaningless statements. The statement "colorless green ideas sleep furiously" is grammatically correct but meaningless. Programs can also be syntactically correct although meaningless. Such programs will not halt because they don't solve any problem. Turing's Halting Problem is the claim that we cannot know when a syntactically correct program is also meaningful, because if we could distinguish between meaningful and meaningless programs, then we would distinguish between the ones that will halt and the ones that will not. All meaningful programs will eventually halt since being meaningful implies being finite (although finiteness doesn't entail meaning). The requirement for meaning is therefore stronger than that for finiteness. Finiteness is a necessary but an insufficient condition for meaning. Current computer languages do not allow semantics apart from syntax. Therefore, machines cannot design algorithms for specific problems, and they cannot know if an algorithm solves a meaningful problem. It is humans who must design algorithms and assess their effectiveness.

As we noted above, we can distinguish between two separate issues: (a) that a problem is meaningful and will therefore halt, and (b) that the meaning of the program was the intended solution. Intention is also meaning, but it is *human* in nature: we want to solve certain problems; they may be problems for us. This second type of meaning is important, but for now we can concern ourselves purely with the first problem of meaning. As we have discussed earlier, meanings are not necessarily universal. There are also personal and contextual meanings; whether a program solves an intended problem lies in the personal space: only some people may want to solve certain problems, as their chosen goal. Since choice is involved in making something intentionally meaningful, it becomes personal, and hence can be separated from the universalist issue that is directly involved with the question of program correctness.

A problem is an *intent*, and a program is a sequence of *verbs*. Nouns define objects to be created and verbs define the steps by which they can be created. Every algorithm begins in some state and ends in another. The Halting Problem says that we cannot validate if a set of verbs connects the start state with an end state, or if a program constitutes a valid trajectory between states, without running it (whether the end state is the intended one). If there is an algorithmic procedure to validate that a given sequence of actions halts—without executing the program—then such an algorithm will tell us in advance whether the program will meet some goal (whether it is the intended goal). We could then separately validate if the achieved goal is the intended goal. The program that decides if a program will indeed halt can be called the Halting Algorithm.

The Halting Algorithm has significant consequences. One result of having such a procedure is that we would not need to manually design specific algorithms for solving individual problems. Given a certain problem, the Halting Algorithm will iterate and mutate over randomly created algorithms until one of such randomly generated algorithms is found to be suitable for solving the given problem. The Halting Algorithm would obviate the need for humans to solve problems since machines would suffice. Turing's proof shows that the Halting Algorithm cannot exist for Turing Machines. Alternately, the human capacity to solve problems cannot be converted into a Turing algorithm. A consequence of that conclusion is that the problem-solving human mind does not work like an algorithmic Turing Machine. It follows that humans must always design algorithms, although Turing machines can execute them. Whether the algorithm will terminate and would have solved the intended problem when it terminates, can only be decided either by executing the program or by a human examining the program contents[12].

The Halting Problem implies that programs must be manually tested. There cannot be a program that inputs software and outputs the verdict on whether the software halts. If this were possible, the software testing industry would not exist, because there would be programs that automatically validate software. Since there cannot be a method for automating algorithm development, computers also cannot automate software development. Both software development and testing must therefore be done manually. Not many people who write science fiction where a machine in a surge of intelligence overpowers its creators give much consideration to this limitation of computer intelligence. A computer (according to the Halting Problem) can only follow instructions. It cannot understand a problem; it cannot create a solution or validate a solution against a problem. The correctness of the program vis-à-vis the intended problem to be solved rests with the human programmer.

An algorithm is a trajectory from start to finish. The start and finish points, on the other hand, define the problem. The language needed to describe the start and the finish is the language of problems. The language needed to describe intermediate steps is the language of solutions. The import of the Halting Problem is that problems and solutions cannot be described in the same way; we need two languages to describe problems and their solutions. The language of problems involves *nouns* and that of solutions involves *verbs*. Current computers do not understand nouns or verbs. A computer does not know if it is flying a plane or operating a robot. Similarly, the computer does not know if the program will end. Like Gödel's theorem deals in names and concepts though they cannot be understood in mathematics, the Halting Problem pertains to verbs producing a noun, but this can't be understood by computers.

To know if a sequence of instructions produces the desired verbs and if the sequence of verbs produces some nouns, we must run the program. So, we cannot know if an algorithm will do what we want it to do, and if indeed it will solve *any* problem, and by consequence if it is indeed going to solve the intended problem.

The Halting Problem can be solved if there was a way to *prevent* the formation of meaningless programs. Since the meaning of a program is the algorithm, we need to find methods by which only meaningful algorithms will be allowed. Once the algorithm is meaningful, it will halt. Once it halts, we can determine whether it solves the intended problem. The crux of the Halting Problem is detecting if a program's algorithm is meaningful. This problem can be further reduced to the need to detect *loops* in algorithms. If an algorithm does not loop indefinitely, then it will eventually halt.

What is a loop? A loop is a set of instructions whose end is the beginning of the same instruction set. Gödel's problem arises when a statement not-P is mapped to P. Similarly, the Halting Problem arises when the end of the instruction set is mapped to its beginning. Just as Gödel's problem can be solved by preventing the mapping of not-P to P, the Halting Problem could be solved by preventing loops.

Every program that halts is equivalent to a tree, and every node in a tree has only one parent. Program semantics is equivalent to the idea that all meaningful programs must have a loop-free tree structure. A tree is a hierarchical structure and I will later show how questions of program semantics also lead us to hierarchical notions of space and time (descriptive semantics, as already mentioned above, also requires hierarchical notions about space and time).

The Structure of Turing's Proof

However, before we get into how the Halting Problem can be solved, let us look at Turing's proof and its associated difficulties. Turing proved the Halting Problem by assuming a converse and then showing that the assumption is logically contradictory. Let us assume that a program that solves the Halting Problem called HALT exists that accepts as input a program P to determine whether P will halt.

```
HALT (P) = 1, if P halts
HALT (P) = 0, if P loops
```

Assume also that we can formulate another program—TLAH—that does the reverse of HALT. If HALT says that P stops, then TLAH will loop forever. If HALT says P loops forever, then TLAH will stop.

```
TLAH (P) {
        if (HALT (P)) { Loop } else { Stop }
}
```

A program is a number and can be passed as an input to itself. In Turing's proof, a program and its input are both represented by the same number. This is achieved by passing TLAH as input to itself.

```
TLAH (TLAH) {
        if (HALT (TLAH)) { Loop } else { Stop }
}
```

There are two instances of TLAH here—the program and the input to that program. We might call these P_TLAH and I_TLAH. The following two cases are possible with respect to the outcome.

```
Case 1: If I_TLAH loops, then P_TLAH stops
Case 2: If I_TLAH stops, then P_TLAH loops
```

Since I_TLAH and P_TLAH are identical, both the above alternatives imply a logical contradiction. Turing takes this to mean that the Halting Problem cannot be solved because the assumption that the problem can be solved leads to logically inconsistent results.

Loop-Free Computation

Note how the idea of looping is so essential to Turing's proof. While Turing injects it specifically as a program instruction, it is possible that these loops could exist unintentionally. While Turing Machines operate on a single instruction and data at any time—as everything in a Turing Machine is written on a tape that is read serially—loops can occur across multiple instructions. For instance, below is a small loopy program that requires at least two instructions:

```
INSTRUCTION 1: GO TO INSTRUCTION 2
INSTRUCTION 2: GO TO INSTRUCTION 1
```

If we were to visualize this program as a flow-chart, we will see a loop. However, if you only look at single instructions at a time, the loop cannot be detected. Turing Machines look at instructions one at a time and loop detection is possible only if we examine the whole program. Human programmers can detect loops in programs as humans can observe the entire program. We don't need to execute instructions to see that there is a loop. A casual look at the above two instructions will tell us that there is a loop, and that's because we have *visualization* capabilities by which we can see *relations* between multiple instructions within a program. This visualization helps us to draw flow charts with a start and an end. If the start and the end in a program are the same, we have a loop. A Turing Machine has no visualization capability because it isn't designed to understand relations between instructions. Loops are not properties of single instructions. They are properties of *collections* of instructions. A Turing machine cannot know if a program will halt by looking at a single instruction. It must look at *relations* between instructions. In the above two-step program, a loop can be detected by analyzing the relation between the instructions. But this problem gets increasingly harder as the number of instructions grow. For instance, in a program with N instructions, the total number of relations is:

$$\frac{N!}{1!\,(N-1)!} + \frac{N!}{2!\,(N-2)!} + \cdots + \frac{N!}{(N-1)!\,1!} = 2^N$$

With only a 1000 instruction program, the total number of relations will exceed the number of atoms in the universe[13]. For the sake of comparison, an average PC has a processor that runs at 1 GHz. This processor can run the 1000 instruction program in 1 us (10^{-6} seconds). It would therefore seem that running the program is much more optimal than trying to programmatically find if it will stop! But that is true only if the program is loop-free. With loops, those 1000 instructions could execute infinitely and thus never halt.

To build loop-free programs, there is a need for methods by which loops can be prevented. Like Gödel's problem can be prevented by preventing the P to not-P mapping, Turing's problem can be prevented by preventing the calling of a function within a function. A program loop is the descriptive analogue of a set that includes itself. To prevent a loop, there must exist the ability to prevent self-inclusion in programs. Self-inclusion (or recursion) is a very fundamental technique in modern computer science in which a program function invokes the same function within itself, but generally with a different *value*. For example, the Fibonacci sequence of numbers is defined by the following recursive technique:

$$F_n = F_{n-1} + F_{n-2}$$
$$F_1 = 1$$
$$F_2 = 1$$

Now, it is possible to write a program Fib which invokes itself although with a different input value on each iteration.

```
Fib (n):= Fib (n-1) + Fib (n-2)
          Fib (1) = 1
          Fib (2) = 1
```

From a computational standpoint, Fib (n), Fib ($n-1$) and Fib ($n-2$) are different Turing Machines because a Turing Machine (TM) is the combination of instructions together with input conditions. But in practice, programs and input conditions can be separated; the input can be passed to a program in real-time, thereby *invoking* a new TM from within the TM. This invocation is possible because the TM in practice runs on a Universal Turing Machine (UTM or a 'computer') which can run any arbitrary TM. The UTM does not know whether some TM invocation is a valid operation or an invalid loop.

The genesis of recursion lies in the idea of a *stored program* where the earlier and later instructions exist at the same *time* in a computer. Since the earlier and later instructions exist concurrently, it is possible for a later instruction to jump to an earlier instruction, thereby creating a loop. This creates the possibility of a finite program that runs for an infinite time. If programs were *real-time* constructs, such that the past instructions are destroyed after being executed, a future instruction could not refer to the past instruction. An infinite program will now require an infinite time to write and could hence never be produced. If we wanted to run the same program again and again, we would have to invoke the same program over and over, but each time the program will terminate. The successive invocations of a program would not be automatic; they would rather depend upon a *choice*. Of course, this choice can be rational and can be automated; for instance, it could determine if the program execution is progressing, regressing or being static.

If the program is a real-time construct, then the only idea necessary to know if a program will halt is if it has a finite size. If the program has a finite size and its instructions are destroyed after execution, then it will always halt. This addresses the first aspect of the question about program semantics, where a program halting is a precondition for it being meaningful. This tells us that a program has *some* meaning, without telling us what that meaning *is*. The second aspect of the semantic question is whether the program solves the *intended* problem. Even if the program is finite, and a real-time construct, we cannot know if it will solve the intended problem without executing it. This implies, for instance, that a computer cannot know if a program is malicious; whether it would do things that we want it to do or whether it will produce outcomes that we do not want. If we must execute the program to find out whether it is malicious, the damage is done by the time we discover the outcome. Program intentionality—i.e. the question if a program is secure or malicious—is a semantic question and it cannot be answered even if we can answer whether a program is finite and will halt.

To have loop-free programs it is enough to allow only those programs that have a tree structure, although this will not tell us if a program is malicious. To know that a program is not malicious, the tree must be semantic rather than physical. The intent of the entire program could be known simply by examining the meaning of the root of the tree. Thus, simply the detection of a physical tree will solve the first part of semantics (the Halting Problem) while the semantic tree will solve the second problem as well (program goal). Detection of program semantics and its various manifestations requires a semantic and hierarchical notion of space and time, as I will elaborate later. Hierarchy prevents loopy programs and semantic hierarchy can help us detect if the finite program is useful or malicious.

Understanding Program Semantics

Every program is a trajectory in a state space that connects points of initial and final states. The gap between the initial and the final states represents a *problem* and the trajectory connecting these states represents a *solution*. Given a pair of initial and final states, there are several possible trajectories that connect them. All such trajectories solve the problem, but they might use different procedures.

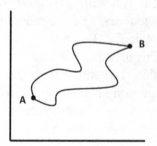

Figure-4 Problems versus Programs

In Figure-4 the gap between locations A and B is a problem and trajectories between A and B are programs. If the state space is *flat*, then it can have loops. But if the state space were hierarchical, it will exclude loops. For instance, in a hierarchical state space, a program could either ascend or descend a path on the tree. If that tree is finite, then the program will always come to a halt—it will either reach the root or the leaf—which constitute program termination points.

In a flat space, there are many paths that join to end points; they solve the same problem in different ways. Therefore, there is no strict association between the program's purpose and the method by which it is solved. The various paths that join two end-points have the same meaning, although they are expressed differently. At best, we could say that some programs are more optimal than others (the optimal programs will follow the shortest path between two points). Even in a hierarchical space, it is possible to construct many paths between two end-points. For example, the path from one leaf to another can pass via higher level branches that aren't in the path. If they do not follow the shortest path, they will have to *retrace* the path from which they deviated, and if we could visualize the path a program takes in a hierarchical space, we could optimize it.

Computer semantics requires a hierarchical view of the state space—drawn as a tree. The nodes on this tree are the states, and the branches of this tree are actions. We can also call these respectively the nouns and the verbs. We might, however, note an important difference between two kinds of verbs—(a) those that *construct* the tree or the object/state, and (b) those that represent the *actions* of the constructed object upon other objects. The latter is also called 'causality' in modern science, and the causal actions are also verbs, but they are different from the *process* by which the object was created. For instance, the causal effect of an object can involve detaching some branches from the tree (along with the twigs and leaves) and attaching them to another tree. The various ways in which a tree can be reduced (while preserving the tree structure) and the removed parts can then be attached to other trees constitute the second kind of verb which we normally term as 'causality'. It is different from the process by which a tree was originally constructed, although causality can also be the cause of tree construction. Therefore, the distinction is somewhat subtle; we can say that the tree structure at any given moment in time is the nodes and their connections; these respectively represent nouns and verbs. If the tree were modified, a new set of nouns and verbs can be created.

Keeping aside these nuances of inter-tree interaction (which needs an entirely separate discussion), we can presume for the moment the existence of a single hierarchical state space and a program as something that traverses up or down in this space. If the *goal* of the program—i.e. the initial and final states—are well-defined, then by the fact that it is possible to measure the proximity or distance in this space, there is a definite process by which a program can determine if it is getting closer or farther from the goal. Therefore, even though loops are possible even in the hierarchical space, it is only if we neglect the *distance* between nodes and have the program do a random walk on the tree. If the program measures the distance on this tree, it can always find the shortest path to the destination. Thereby, we not only avoid the problem of program semantics (i.e. whether the program halts), but also the intentional problem (i.e. whether the program meets the intended goal).

We saw earlier how Gödel's problem can be resolved by a semantic notion of space in which names and meanings are identical; a statement with the name P can never have a meaning not-P. Program semantics needs to extend this notion to include a semantic difference or the notion that the proximity or distance between these points represents the action by which these names and meanings can be constructed. To go from one state to another, we must perform an action. The meaning of the locations denotes descriptions of the state, while the meaning of the distances represents the actions.

Issues in Turing's Argument

The relation between the Halting Problem and program semantics allows us to ask the following question: Can we say that a program is meaningful without knowing its meaning? This question is important because Turing treats the Halting Problem as being independent of the problem of semantics. The HALT program determines if a program P will stop (which means that P is meaningful and is expressed as a tree) without worrying about program P's meaning (i.e. if the tree is semantic). The problem here is that if the tree is non-semantic, we cannot determine the goal it is trying to accomplish (i.e. the output of the program) and without knowing the goal, the program cannot determine if its distance to the goal is reducing or increasing. If the proximity to the goal cannot be determined, then the program can continue looping—even if the tree were physical. So, the key criterion that avoids looping is the ability to determine if the program is getting closer to the goal, which entails that we should be able to solve both the semantic problems—(a) the program halts, and (b) it achieves its intended goal. The program must have an intended goal in the form of the output it is expected to achieve, because without it there is no way of avoiding the program looping.

We cannot determine if a program is a tree unless the tree is also semantic. A semantic program will automatically be a tree, although the tree is not necessarily semantic. Loops cannot be detected in a non-semantic tree without traversing the tree (which amounts to program execution). Thus, it is impossible to know if a program will halt unless the program is semantically constructed. If the program is semantically constructed, then it will always halt. The key question now is not whether the program will halt but whether it meets the intended purposes (or if it is unintended or malicious).

Thus, we can't say that a program P is meaningful (that it will halt) without knowing its meaning (the state it will construct upon halting). To know that a train will halt is to know the station it will halt at. If we can't define the end station of the halt, we can't claim that it will halt. Therefore, despite Turing's contention, the Halting Problem cannot even be stated without a semantic formulation.

The semantic problem is obviously stronger than the Halting Problem, because the semantic problem requires us to know the end-state besides knowing that the program has some unknown end-state. The Halting Problem only concerns itself with the question of whether a program has an end-state, without asking what that end-state is. If we can solve the intentional problem, then we will also solve the Halting Problem, although the reverse is not true. Turing's proof can now be interpreted as the claim that the Halting Problem—which pertains to knowing whether a program will halt without knowing its meaning—can't be solved. That is, if we don't know a program's intention, we also cannot know if it will halt but if we know a program's intention we will also know that it will halt. There may be programs that halt and produce a meaningful output, but an automated procedure can never determine that. Humans will have to decide that—by examining the program, or by executing it.

The flaw in Turing's argument is the assumption that we can pass a program as input to another program, which Turing postulated as HALT (P). Obviously, HALT is not expected to execute P, so what can HALT do with P? Of course, HALT can compile the program, which will tell HALT whether the program is syntactically correct. But syntactical correctness does not guarantee that the program will halt. The notion that we are passing P to HALT is incoherent because P is a program and there is no way in which HALT can use this input in any way other than to execute P, which HALT is not supposed to. The proof assumes that we can pass P as an *input* to HALT, in a way that is not program instructions, without defining what that scheme is. From above, P can be passed to a program as an intentional description of a program. If we are passing the meaning of P to HALT, P is already meaningful. And if the program is meaningful, then it is guaranteed to halt. HALT now effectively does nothing; it receives a meaning and confirms it is meaningful.

Thus, if we can pass P's intention to HALT then P halts and HALT shall return 1. By implication, TLAH which is the inverse of HALT shall always loop. This infinitely looping TLAH will never construct any object and can therefore never be described intentionally. TLAH can therefore not be passed as input to any program (including itself). The key to Turing's proof—TLAH (TLAH)—cannot intentionally exist, and the logical contradiction cannot be constructed.

Both HALT and TLAH are used in incoherent ways in Turing's proof. A program P (not its intention) is passed to HALT, which is incoherent because HALT is not expected to execute the program. Then TLAH, which always loops (if P halts), is passed as input to itself, even though TLAH cannot be represented as data input. Turing's proof doesn't make the Halting Problem unsolvable, because a program's halting is decided if the program has a useful purpose. We don't need a HALT program to know if a program P will halt. Every program will halt if it is meaningful. We only need to validate if the program's meaning is not unintentional or malicious.

3

Mathematics and Reality

As far as the laws of mathematics refer to reality, they are not certain; as far as they are certain, they do not refer to reality.
—*Albert Einstein*

The Dogma in Mathematics

A common presupposition amongst mathematicians is that their craft pertains to a Platonic world of pure forms or that mathematics deals with pure ideas which are imperfectly reflected in the present world of things. When these pure forms are reflected in matter, then matter can be described by mathematical theories. The world of mathematics is, however, a world of possibilities and only some parts of the possibilities are actualized in the present world. The mathematician therefore studies what is possible, not what is real. This view of mathematics fails to explain how the Platonic world is reflected in the real world. The real world includes two things. First, it includes the mind of the mathematician, which grasps the ideas in the Platonic world. How does the mind know the Platonic world? Second, it includes material objects that conform to the mathematical ideas. How are the ideas in the Platonic world reflected in the real world? Finally, even if the ideas in the Platonic world are reflected in the world of things and the mind of mathematicians, how does the mind grasp these ideas from the real world rather than the Platonic world? This is called the mind-body interaction problem.

Mathematicians use minds to create mathematics, but they neglect the problem of what the existence of minds brings to their assumptions about mathematics. This oblivion is convenient: we can do mathematics without worrying about reality or the mind. But, to the extent that we disregard this problem, our mathematics is also incomplete or inconsistent, because ordinary language—as opposed to mathematics—is not Platonic. Ordinary language supports more distinctions (for example, the distinction between the physical, the conceptual, a program and a purpose) than mathematics. Without incorporating these distinctions, mechanical symbol manipulation cannot be consistent and complete, and it cannot therefore explain the conceptual, reasoning, and intellectual capabilities in the human mind. To explain these, we will have to formulate a mathematics that supports the categorical distinctions in ordinary language.

The difference in mathematics with respect to ordinary language brings two problems to mathematics. First, logical systems cannot deal with the meanings of statements and programs. Second, since mathematics cannot explain the meaning-carrying power in the minds, it cannot be used in the scientific description of minds, requiring the invention of another language to deal with minds. Both these problems are related to meaning; the first problem pertains to meanings in formal systems (including machines) while the second to the existence of meaning in living beings (e.g., humans).

These two problems suggest an alternative to Platonism. In this alternative, ideas, minds and things are not in separate worlds. Rather, they exist in the same world. The world of things is therefore a world of symbols and not a world of *a priori* independent objects. However, we can separate them into three kinds of languages—universalist, personal, and contextual. The universalist language deals in pure ideas. The contextual language converts these universal ideas into individuals and places them in relation to each other—which may be a different placement than that between the ideas themselves. For example, in the universalist conceptual tree, the ideas 'table' and 'chair' are not adjacent. But tables and chairs can be adjacent in the physical world, creating a study table and chair. Their proximity creates an additional property, which was absent in the pure concept of a table and a chair—namely, they are a now 'study' table and chair. Similarly, the universalist ideas can appear in the mind as individuals, but the different levels of focus, attention, and importance we give to these ideas varies from person to person. Some of these ideas are more relevant than others; that relevance is a personal property and drives people in different directions.

Therefore, the pure world of ideas is within this world, but it is only one of the three aspects of this world. It identifies a tree of conceptual states and actions. Additionally, there is a world of individual objects which *instantiate* these ideas into objects and place them into context or relation to other such objects. These two worlds cover the 'inanimate' world, which doesn't have personal intentions or goals. As we have seen, this inanimate world is incomplete in solving the Halting Problem, because a machine can loop indefinitely if it doesn't have a goal, and without that goal, we cannot determine if it is meaningful (only humans can decide if it is meaningful). This ability in humans to determine if the program is meaningful identifies another type of world which exists internally, but it measures proximity and distance in relation to the *self*. The things that are meaningful to me are close, and those that are meaningless are far. This near and far is regardless of the universalist or contextual distance; therefore, the personal world is a separate world.

In practice, each of these three worlds are combined. For example, we observe the world of individual objects, and see them as distinct things. But then we also conceptualize these objects in terms of ideas and form scientific theories of nature. Once these theories have been formed, we construct technological products that are useful to us. If therefore you observe a machine, it has the universalist ideas, it is an individual object, and it has a useful purpose. The combination of these worlds makes the object meaningful.

Moving away from Platonism has distinct advantages for mathematics. By accepting that the ideal world of mathematics and the real world of objects are present in the same things, we realize that the universal, the individual, and the intention are co-existing. These are of course logically distinct, and therefore, they can be treated as being different domains or 'worlds', but they overlap everywhere. Their logical distinctness allows us to introduce modes in language, but their combination prevents us from separating these modalities into different realms. Furthermore, each of these three domains is semantic, although in three different ways—universally, contextually, and personally. Thus, the logical separation followed by their combination produces the everyday lived world in which mathematics is relevant because it is one of the three worlds.

We can identify these worlds as 'mind' (which deals with personal goals and intentions), 'matter' (which deals with contextual relations between individual objects), and 'mathematics' (the universalist realm of pure ideas and concepts). The 'real' world that we observe, experience, understand, theorize, and use, combines the three worlds, so they can be talked about separately, but they cannot be divorced from each other. Specifically, we must treat the world of objects as *symbols*, which have a universalist meaning, a contextual relation to other symbols, and a personal significance assigned by different observers. This requires the ability to include multiple ordinary language categories (names, concepts, algorithms, problems) into mathematics, and the ability to derive them from the *same* symbol. The consistency of number theory implies that numbers that will denote names, concepts, programs and problems are different types of numbers. The completeness of number theory implies that although names, concepts, programs and problems are different categories, they can be used interchangeably. This is because every problem can be solved, every concept can be instantiated into an object, and every object has a meaning. In one case you may be referring to a goal, but what you really mean is the attainment of an object that solves the problem. In another context, you may be speaking about an individual object, but to understand it you must invoke the universalist concepts in a theory. Meeting these two requirements needs a shift in our notions of space and time. Locations and directions in space-time are now names by which objects are individuated but they also denote meanings[14].

The Platonic notion of mathematics is ultimately detrimental to mathematics because questions of meaning and purpose must be solved within mathematics, if mathematics must describe the thinking and intentional capabilities and must solve problems of incompleteness and inconsistency arising from a Platonic view. Ordinary languages can express these capabilities, but mathematics cannot without creating paradoxes. Since physical sciences use the mathematical theories—which themselves cannot deal in meaning and purpose—how could these theories *in principle* describe the human semantic and intentional capabilities? In fact, given the problems of meaning in current mathematics, no theory of matter that employs numbers can even *represent* the meanings, let alone explain the creation and exchange of these through matter.

Most physicists who aim to provide an explanation of everything in the universe—including the existence of consciousness—seem not to give much thought to this problem. If nature is mathematical in the sense that current mathematics describes it, then no physical theory can ever explain the mind, not because we will not find some physical phenomena corresponding to the mind, but because even if we did find it, we will not have the mathematics to describe it. Ideas about mind and meaning are everywhere. They are not something hidden from us, or something that we need to 'discover'. And yet, we don't have a way to *describe* or *represent* them in mathematics. What physical sciences discover would then be presented inadequately, and the incompleteness of mathematics will trickle into the other sciences. If mathematics cannot represent meanings without creating contradictions, then a physical theory about mind could never be mathematically formulated. Therefore, just searching for the mind in the world is useless, unless we have a language that can present its properties. This makes mathematical incompleteness a fundamental stepping stone to fixing the incompleteness in all of science.

This approach to mathematics requires us to discard Platonism. We are not in a pure world of ideas as mathematicians believe, nor are we in a pure world of things as physicists believe. We are in a world that consists of both ideas and things. Mathematics needs a new approach in which ideas and things can co-exist without creating contradictions. Such an approach was attempted after Plato by Aristotle. Let us now take a closer look at the Aristotelian solution.

Hylomorphism

Platonism was not accepted by everyone even during Greek times. Opposing Plato who believed that *form* resides in another world of forms, Aristotle argued that forms are immanent in objects, which combine both matter and form. Aristotle's theory was later termed

Hylomorphism, a combination of two terms—*hyle* (matter) and *morph* (form). Of course, Aristotle did not put forward an explanation of how matter and form combine. Yet, he traced the need for this combination to some problems in Platonism, which he addresses through the combination of matter and form. These problems must be solved in any theory that uses both meaning and matter.

The key problem with Platonism is the cognition of ideas—how do we intuit a new idea or theory? If there is another world of pure ideas, as Plato claimed, how do we connect to that world? The other-worldliness of Platonism is not just a philosophical problem. It is also an empirical problem since it contradicts the process by which newborns acquire ideas like table, chair, book etc. Ideas come to us during childhood through repeated acquaintance with things, not via a perfect world of ideas. Most of our ideas about the world are generalizations of what we see in the world and we are taught early in childhood to learn to convert specific things into universal ideas which can be applied even outside the contexts where we learnt them originally. But generalizations have a problem in that they make ideas secondary to the physical world. After all, if ideas are generalizations of the physical world, then the world must exist before ideas do. Perhaps cognizant of this problem, Plato took the reverse approach. Instead of saying that ideas come from the world, he says that the world comes from ideas. Plato believed that behind the façade of appearances, there is a real world of pure ideas which is reflected in the appearances. The unsolved problem with this view is how the Platonic world is reflected in minds and in matter.

To address these problems, Aristotle said that the world of ideas is the world of things because objects are comprised of *matter* and *form*. Matter is physical, and form is conceptual. Matter is amorphous, and form combines with it to give it physical properties like shape and size. Thus, the *shape* of a pot is due to its *form* and not due to matter. This leads to the problem that if the *form* of the pot is its physical *shape* then *potness* must be defined as a specific kind of shape. And it is here that cracks begin to develop in the theory.

If a pot is a specific kind of shape, then objects with different shapes cannot be pots. Even more importantly, the symbol of the pot will also have to have that shape to denote the idea of pot. To address these problems, the notion of *form* must be defined in many ways. The shape of the pot is not the only reason we call something a pot. If things can be used as pots, then they are functionally a pot, even though they may not be identical to the shape of an ideal pot. The shape itself is given relationally between different parts, and if the relation—i.e. the functionality is satisfied—the thing in question is a pot. Similarly, a symbol of a pot doesn't have to look like a pot if in a context of symbols, the relation between the symbols can denote the idea of a pot. Unlike imperfect pots, which have some similarity to the shape of a pot, the symbol of a pot may not even look like a pot and still denote it. The meaning in the symbol is not necessarily in the shape but could also exist in the relation between the symbols by their use. Thus, while imperfect pots have some similarity in their shape, and the meaning is associated with the shape, in the case of symbols, the meaning can be through relations, not shape.

Even the Platonic notion that every concept is associated with a pure form that denotes the perfection of that concept—independent of other concepts—is now false. Even perfect forms must be defined in relation to other forms, and we can grasp a system of concepts collectively, never concepts individually. This implies that ideas are grasped through a hierarchical relation between more and less abstract ideas, rather than just as pure forms, which are individual shapes. We cannot know a single idea by itself. We can only know the ideas through the differences between ideas. For instance, we cannot know the idea of 'yellow' without knowing how it differs from 'red' and 'blue'. Similarly, we cannot know what the difference between 'red' and 'yellow' unless we formulate a higher idea 'color'.

Plato and Aristotle did not address the personalist approach to meaning, and this neglect was addressed only in 20th century Continental Philosophy. To the extent that Continental Philosophy (practiced and preached largely in mainland Europe) has not adequately intermixed with Analytic Philosophy (practiced and preached largely in the English-speaking world), the conclusions of one remain divorced from and opposed to those of the other. To understand meaning, we now need to use three kinds of approaches—universalist, contextual, and personal—and combine them. I will develop this theme in the subsequent chapters.

After Aristotle, Greeks equated form with shape, which implies that concepts could be derived from physical percepts. This idea was further extended during 20th century Positivism, which claimed that concepts have no separate existence; they are merely concise labels for a collection of properties. The Platonic world of ideas thus first descended into matter in the hands of Aristotle and was then eliminated from matter in the hands of the Positivists. The net result of such an exercise is that we lose the ability to speak of concepts, and thereby the ability to encode concepts in matter. Equating forms with shapes thus creates many problems and a new way of understanding concepts is needed. As we can see, the Greek notion of form—as something that stands on its own—is incorrect. Even pure concepts are to be understood in relation to more abstract concepts, thereby constructing a tree or hierarchy of ideas. The meaning of forms therefore comes not purely by their forms (although form is important) but also through the relationship between forms. In fact, these relations exist not just between ideas, but also between things. Therefore, some objects can *contextually* become representations of meanings if the relationship between them has similarities to the relations between the pure ideas. We can also culturally designate certain symbols with meaning in certain chosen contexts.

The relational view makes it possible for us to think of perfect pots as well as things that may not be ideal pots but could be thought of as a pot in relation to other things through a contrast or distinction. It also makes it possible to understand how symbols that don't have the shape of a pot could denote potness. Basically, if a symbol encodes the same distinctions with other symbols, as they exist between pots and other objects, then the symbol can denote potness.

Computation and Motion

Semantic notions are relevant not just to pure ideas but also to computation. Early notions about computation were derived from the physical idea of motion. For instance, Charles Babbage built a machine using gears and wheels to compute numbers. While modern digital computers have gone quite a distance from the classical idea of computation (specifically in the sense that modern computers are no longer computing through moving parts), the basic underlying idea of computation through of physical change has not been altered. In effect, we do not recognize computation as a fundamental category in nature, just as we do not recognize concepts as a basic category. The problem with treating computation as an *emergent* rather than fundamental property of nature is that physical change is itself a byproduct of the computation of natural laws. Therefore, if computation is a product of physical change and that change in turn depends on the computation of laws in physics, then we have a circular dependence between change and computation.

This circular dependence between change and computation can be used to create another kind of logical paradox. We could use a computer to construct a statement that denies change. This construction need not be logically derived from axioms. We might also freely construct such a statement using a computer that denies change. The statement's existence is now the following logical paradox: Given that computation depends on change, if change did not exist then the statement could not have been computed in the machine. The existence of the statement can be taken to be its falsity which implies that change must exist in nature. On the other hand, if change exists then a computer can also compute the statement that denies change. The computability of the statement can be taken to be its truth and hence a denial of change. In other words, if change exists then it does not exist, and if it does not exist then it exists; the existence of change is logically contradictory if the existence of change depends on the mathematical computation of laws.

This problem has an intuitive everyday counterpart. Assume that a child denies the existence of her mother. If we take what the child says to be true, then the denial means that the child itself could not exist and the truth of the statement would imply its falsification. But if the child is lying, then we have the problem of how the truth of the mother's existence generated a lie in the form of her child's falsities. This problem has a solution: there is a difference between the meaning and the truth of statements. A child can deny the existence of its mother, but the existence of the child does not entail (a) that such statements cannot be made or (b) that because they can be made they must necessarily be true. Note that the denial of change is a semantic problem and not a syntactical one. Modern day computers can produce a statement that denies computation without ever realizing that they incurred a contradiction. But if we think of a computer as a formal logical system, such a system cannot produce a proposition not-P beginning with axiom P. In a computer, this is quite trivial: we only apply the *not* operation on the axiom P to produce a proposition not-P. Why is the *not* operation valid in computation when it is not a valid operation in logical deduction?

The simplest answer to this quandary is that the not operation reflects our everyday capabilities to create new ideas by negating existing ideas. The idea and its negation represent the simplest type of two-way *distinction*. Everyday concepts are defined through such oppositions. For instance, the idea of 'black' is defined in opposition to 'white', and 'everything' is defined in opposition to 'nothing'. Thus, 'white' can be used to represent the idea of 'everything' and 'black' the idea of 'nothing', because they encode a similar form of mutual opposition. The foundations of our language are not in a mutually consistent set of ideas. They are rather in a logically opposed set of distinctions. Indeed, if the world has meaning, then it also must have the opposites—which is how meanings are defined. If we begin with a foundation of mathematics in a mutually consistent set of ideas, then such a foundation should never produce a contradiction. That would in turn mean that our world could only have one side of the distinctions—e.g. hot but not cold, bitter but not sweet, sad but not happy, etc. That clearly contradicts the reality of the world. On the other hand, if we begin with the idea that the foundation of mathematics lies in a basic set of logical distinctions, then any idea—even logically contradictory ones—which can be produced by combining these basic distinctions would be meaningful propositions in mathematics. Their 'proof' is simply the succession of steps by which they were produced. But their 'truth' can only be known through a *consistency* amongst a wider set of propositions.

The remarkable insight from here is the need to separate *proof* from *truth*. A proof represents a succession of steps that produce an output. A computer can produce a logically contradictory statement, and we cannot deny that the computer proved such a statement. But that proof does not necessarily imply that the statement is true.

Thus, when a child denies the existence of its mother, the statement can be logically *proved* although it is not *true*. We now have a variation of the Gödel's paradox where we can prove falsities. While this seems confusing, it is worth noting that it is made possible by allowing contradictory axioms as distinctions and by permitting negations in the act of computation. The process by which such a statement is produced can be codified as a program. The program represents a set of logical steps of combining and mutating axioms, but the *choice* of steps and the order between the steps is not logical. That is, the operations performed by a program are a set of logical tools available to us, but whether we choose to use the tools, and the order in which they are chosen, is not logically dictated by the axioms. Through this choice, each program represents *new* information which is not logically implied from the axioms, although it can be produced through an appropriately codified program. The program is now the process by which new information can be produced. We can also say that new information is produced through programs.

If mathematics must explain the ability in real systems to produce contradictions, then it must change its foundations. The language underlying mathematics must employ mutually opposed distinctions rather than mutually consistent ideas. This will allow mathematics to represent meanings, since meanings are always given in relational opposition to other ideas. Further, when we allow a foundation of distinctions, we will allow the possibility of logically contradictory ideas. There is no harm now in allowing the negation operation, because a negation operation is simply the other side of some binary distinction. A succession of operations on axioms is now a valid program which can be proved but it not always true.

This takes us to the heart of the question: What is mathematics? Is it a system for finding only the subset of all statements that are mutually consistent? Or, is it also a system for generating all possible statements, even if they are contradictory? This question about the nature of mathematics is also closely related to the nature of reality: Is reality only the things that can exist consistently? Or, is reality also the system for generating all possible things, even if they are contradictory? Note that if reality allows contradictory things but mathematics does not, then we cannot describe that reality using mathematics. If on the other hand reality forbids contradictions but mathematics allows it, then mathematics cannot be used to accurately predict the reality. Current mathematics is the system for generating mutually consistent propositions while reality is clearly the system for producing all statements, even contradictory ones. That indicates a shift in the approach towards mathematics.

What we consider contradiction in current mathematics is also the route by which symbols can denote meanings, because meanings are given via distinctions and oppositions. Permitting distinctions does not rob us of the ability to know when they are contradictory although it allows us to add meanings into mathematics.

Current mathematics is based on the idea that nature is a *consistent* system. Physical theories however postulate laws of nature which are *formulae* that convert initial states to final states. A final state in physics can contradict the initial state; for instance, a particle can move in the opposite direction from its initial state. This is made possible because we employ formulae that transform the initial state. This transformation is not simply the rules of logic. It is rather a *program* which permits the inversion of states. The existence of laws of nature and programs asserts the fact that nature does not permit only logical transformations. It also permits certain laws, which go beyond logic in the specific sense that they permit negation as a valid operation. It seems, therefore, that nature is a computational system and not a logical system. In a logical system, P and not-P *cannot* exist simultaneously. In a computational system, P and not-P *can* exist together. In a semantic system, P and not-P *must* exist together because they are only defined by that distinction.

The Foundation of Knowledge

The theory of knowledge is called epistemology and it discusses the methods by which we can acquire certainty. All attempts at knowing aim to find something fundamental in nature from which other things can be derived. But can we find such fundamental ideas in nature? The problem stems from the fact that our beliefs are in turn based on inferences from some other beliefs which in turn depend on other beliefs. When beliefs depend on other beliefs, there are several possibilities. First, the chain of beliefs never ends, and you can always find something more fundamental, forever. Second, the chain of belief circles back on the original belief. Third, the chain stops and ends in beliefs that cannot be based on something more fundamental and these beliefs can be justified. Fourth, the chain stops in some beliefs that can't be justified. Of these possibilities, the first and the fourth deny that knowledge is possible. In the first case, you cannot know because the foundation of knowledge is a bottomless pit. In the fourth case, you cannot know because the roots of the tree of knowledge end up in the sky—i.e. abruptly and without adequate justification.

The second and third options allow knowledge. The second one is called Coherentism and the third one is called Foundationalism. Coherentism essentially says that you cannot find some fundamental ideas because all ideas circularly justify each other. You cannot say which of these ideas is more fundamental, although all of them collectively are *coherent*. Foundationalism says that knowledge exists as a tree rather than as a circle. There are indeed some fundamental roots from which the trunk and leaves emerge.

There are however at least two ways in which we can conceive Foundationalism—non-semantically and semantically. Current mathematics views knowledge foundationally but non-semantically. That is, knowledge is based on a fundamental set of *consistent* axioms. This view, as we have seen, leads to contradictions when semantics is added to mathematics through names, concepts, algorithms and problems. There is, however, also a semantic method to conceive Foundationalism where the fundamental ingredients of nature are *distinctions.* By their very nature, distinctions are not consistent. And since they are defined relationally, these can be used to represent fundamental ideas *semantically.* The difference between non-semantic and semantic foundations is that the former allows only a consistent world of propositions, but the latter permits contradictions. This is not a *self-contradiction* so mathematics as a proof system is still consistent. However, the provable statements have a denial which cannot be co-located with the assertion.

For instance, assume for the moment that the fundamental types represent a tri-partite distinction like yellow, blue and red. These ideas are defined mutually or not at all. When these types are combined to create more complex types, a diversity of types would be created but they would not be consistent either. However, we can now *classify* these types into sets that are *relatively* more consistent. For instance, there can be classes that consist of different shades of yellow, blue and red, and these can be called classes that encode the meanings yellow, blue and red. These classes will now also depict the mutual distinction with other classes, like the basic distinctions from which they were created. For instance, the class yellow will denote a distinction with the classes blue and red like the distinction between yellow, blue and red. However, when we define complex types by mutating elementary types, then these complex types can never be completely consistent. This is because the very distinction between these types—which represents their individuality and separation—is itself founded on a fundamental distinction or opposition.

In current mathematics and physical sciences, we assume distinct objects without defining how they are distinct. It is assumed that objects are *a priori* individuals, and their individuality is independent of other objects. With such *a priori* individual objects we cannot encode meanings (and the attendant problems of meaning in mathematics and physical sciences follow). All the individual objects are however mutually consistent fundamentally because they are all *defined* to be of the same *type*. For instance, in classical physics, all objects are defined to be particles. All particles are of the same type—they are all particles—and they are hence logically consistent. However, these particles cannot denote meanings because they were of the same *type* to begin with. To denote meanings, each particle or alphabet must be of a different type. These types must be defined though a mutual opposition and not independently. For instance, there can be two particles, one black and another white. When we define the particles through a mutual opposition then each particle has a different type, and that type difference entails that they are all mutually opposed in some way. If we assume *a priori* distinct particles of the same type, then their distinction is physical and not semantic. This distinction cannot denote meanings. To denote meanings the distinction between particles must be semantic, and this implies that all these particles are mutually inconsistent.

Now two particles are different because they are white and black. The distinction between these particles also represents an opposition. And that opposition is also a contradiction.

To hold meanings physical individuation must be a consequence of a type individuation. But this also implies that every object distinction is also an opposition. Now there can never be a consistent set of objects, although these objects will now represent meanings. The problem of meaning goes to the root of the idea of consistency in mathematics. A semantic mathematics cannot be consistent in the sense that we currently construe it because any two objects are different due to being opposed to each other in some manner; if one object denotes white then the other object can denote black. It is because they denote different ideas that they are different entities. If the idea difference is removed, the individuality is also lost. This is a semantic notion of objects because objects now represent ideas. But it is not consistent in the current mathematical sense.

It follows that no collection of ideas is completely consistent, although we can formulate collections in which these inconsistencies are divided into mutually orthogonal ideas, like how we can divide all the shades of yellow into some finite set of yellow shades by combining which other shades of yellow can be formed.

Such collections are *coherent,* but they are not *consistent.* We can say that each such collection represents a set of ideas that are mutually justified—because they are mutually defined—but they represent mutually opposed ideas. Their individuality and meaning is defined as semantic opposition. We have now reconciled the differences between Coherentism and Foundationalism. The foundation is semantic, and it comprises basic distinctions. When these distinctions are aggregated, they can be divided into a mutually opposed set of meanings, which reinforce the other meanings in that set. However, in the process of reconciling Coherentism and Foundationalism we sacrificed *consistency* and, with that, the basis of current mathematics. No proposition in such a mathematics is ever consistent with another proposition because each proposition is distinct from another proposition due to some type distinction. These axioms can only be *understood* by their mutual opposition.

My main aim in the above paragraphs was to illustrate that current mathematics is fundamentally opposed to semantics because it is founded on the idea of consistency. To induct semantics into mathematics we need to discard the idea that mathematics is about deriving consistent propositions from axioms using logic. A semantic mathematics will rather be the enterprise of constructing varied propositions from a fundamental basis of mutually opposed ideas or distinctions. No two propositions in this mathematics will ever be consistent in all respects, although they can be consistent in *most* respects. When two propositions are completely consistent in all respects, they must only represent a single proposition. For instance, they may be two instances of the same proposition. On the other hand, if propositions represent different (non-contradictory) meanings then they can be coherent although not consistent.

Objects Created from Meanings

The semantic viewpoint suggests that if we see two symbols of meaning, they must be *created* from a distinction. The symbols denote different types and this difference is given by the distinction. If we begin with the idea that there are objects that exist independently, then we cannot give these objects a meaning because the independent objects will never become mutually opposed. Even within a collection, the independent objects will remain unrelated. Such objects, therefore, can never represent meanings. To encode meanings, we must discard the idea that there is a world of mutually consistent objects which is then somehow made to encode meanings. We must begin with the idea that there are meanings, which are then used to create mutually distinct meaning-symbols.

In current mathematics, everything begins in the idea of objects. We suppose that when independent objects are collected, the set of these objects represents concepts. In physics, space-time points are derived from the intersection of objects (trajectories)[15] and this idea is called the Dedekind Cut[16] in mathematics. More sophisticated ideas such as order, quantity and structure are also constructed from objects. Numbers are formalizations of order and quantity and algebras are generalizations of numbers. In other words, every mathematical construct is derived from the notion of an *independent* object. This idea of independent objects is antithetical to the idea of meanings because meanings are defined by oppositions and independent objects can never be opposed. This problem can only be addressed in one way—we must *construct* the idea of object from meaning. This would imply that the distinction between objects is not given *a priori* as we presume in current mathematics. Rather, meanings are *a priori*, and these meanings always exist as distinctions. They can be used to create distinct objects. The individuality of the objects, which are constructed from meanings, would be that these objects are symbols of distinct meanings.

It follows that to incorporate meanings in mathematics we cannot assume a distinct world of objects and then derive sets, order, quantity, structure, number and space-time from it. We must instead derive objects from sets, order and space-time. Sets cannot be defined individually; they must be defined in distinction to other sets. Sets and order represent distinctions, which when applied to space-time will create distinct objects. However, since sets and order are logically prior to objects, the sets and the order must be defined in a new way—a semantic way. For instance, we must now define sets as representing concepts, and order as representing the succession of concepts. The difference between concepts and the order between these concepts denotes distinctions. These distinctions represent locations in space and time, and these names are also objects. These objects will not be independent, since we never construct a single object. Rather, objects would be defined by their distinction from other objects, and the difference between objects would be *isomorphic* to the form of the distinction. Thus, a two-way distinction will construct two distinct objects; a three-way distinction will construct three distinct objects, and so forth. Objects constructed thus will be defined by their distinction from other objects and will hence represent the semantic distinction between meanings.

This approach to defining objects has important implications for physical theories. The implication is that we cannot assume that nature is *a priori* distinct objects, which then aggregate to form brains which then somehow encode meanings. The problems of semantics in mathematics show that we cannot construct meanings from objects, although we can construct objects from meanings. A mathematical theory of nature must therefore derive the idea of objects from a new type of reality in which sets, order and concepts are more fundamental constructs than that of the objects.

The problem of meaning represents a revolution in mathematics where many of the fundamental ideas in current mathematics must be reversed. Earlier we saw that semantics requires us to discard the idea that mathematics is about consistency and logical derivation of theorems from axioms. Now we see that even the idea that objects are fundamental is an impediment to semantics. A new foundation of mathematics is needed in which objects are constructed from distinctions and these distinctions are prior to the objects that they construct. In this foundation, the mind will not be derived from objects. Rather, objects will be constructed by the mind. When the mind constructs objects, then objects are defined as different types. The *identity* of these objects is that they are symbols of a different meaning. These symbols are both physically and semantically distinct and these two distinctions are identical (contrast this with current mathematics where symbols are physically distinct, but their semantic distinction is unknown). Such an approach to symbols can be consistent and complete because by knowing the physical individuality of objects we can also know their semantic distinction. For instance, if the location of an object represents its physical identity, then that location also denotes its meaning. We can use the physical and semantic identities interchangeably, and this implies that a proposition not-P can never be given a name P, and paradoxes that arise in Gödel's and Turing's theorems cannot occur now.

The Question of Truth

A semantic approach to mathematics, however, raises an important question: If many propositions are inconsistent, then how do we determine the truth of a proposition? In current mathematics, we determine the truth by deriving a theorem from axioms. However, now, these derivations only represent *proofs* and not *truths*. Every proof is a choice of a program that mutates and combines elementary distinctions. By such proofs we can construct new propositions, but how do we know if these propositions are indeed true?

Since we can construct arbitrary statements through choices, all these statements can *exist*. The classical approach in mathematics, where we regarded theorems true if some object exists, does not work anymore. We must now rely on a new approach indicated by Coherentism. In this approach, the truth of a proposition is relative to other propositions in some collection of propositions. Of course, no proposition is completely consistent with any other statement. But some such statements can be *compatible* with other statements in one of the following three ways. First, some statements can be *descriptions* of other statements. Like a picture encodes a reality and the picture and reality are similar although not identical, some statements (that we call descriptions) can be consistent with other statements (that we call reality). This is the classical empirical notion of truth, except that it can exist within mathematics through a semantic isomorphism between two propositions. Second, some statements can define the *problems* that are solved by some program statements. To solve a problem, there must be an isomorphism between the program and the problem. This is the classical pragmatic notion of truth, where ideas are true if they work or help us achieve goals; however, it can also be defined within mathematics through a semantic isomorphism between problem and program. Third, statements may not solve a problem or be descriptions, and yet collectively they form a collection of statements that may be a description or be useful. This is the sense in which Coherentism is used currently: it is a fact about a collection of statements that they become sensible, form a theory, which can then be used to describe the world, or to build technology that solves problems.

While we lose the classical *consistency* conception of truth when semantics is brought into mathematics, we can incorporate new *empirical, pragmatic* and *collective* notions of truth in mathematics. The consistency notion of truth is that a statement is true in all possible worlds. However, the empirical, pragmatic, and collective notions of truth are contextual. Since knowledge is organized as an inverted tree, this contextual truth becomes absolute as we rise higher in this tree. Therefore, the branches and leaves remain contextually true, while the root becomes true absolutely. In effect, mathematical truth is also real in some part of the world.

The Problem of Infinities

If objects are individuated by distinctions, then to produce two objects that are arbitrarily near (in a conceptual sense) we must employ smaller distinctions. As the distinction gets smaller, the objects—and the points in space-time on which these objects are located—must also get nearer. Eventually, for an infinitesimal object we would need an infinitesimal distinction. This infinitesimal distinction must exist prior to the existence of objects so that it can be used to create infinitesimals. Since the distinction represents a meaning, there must be infinitesimal meanings to produce infinitesimal objects and infinitesimal distances between them.

Constructivists and Intuitionists have opposed the use of infinitesimals in mathematics. The problem is that the *idea* of an infinitesimal is *constructed* by a *limit* which tends towards zero. In practice, this requires us to apply a finite distinction an infinite number of times. Think of a finite binary distinction such as black-white. This distinction represents a finite amount of information. We can divide each side of the distinction by the distinction itself, creating black-black and black-white. If the black-white distinction required one bit of information—e.g., 0 and 1—then each successive application of the distinction will increase the number of bits that are necessary to represent the newly produced objects. For instance, upon the second iteration of the distinction, we will require two bits to represent black-black, black-white, white-black and white-white as 00, 01, 10 and 11. At each successive division, the number of bits needed to denote the object doubles. By the time we reach infinitesimals, the amount of information needed is infinite.

Infinitesimals therefore require the ability to encode an infinite amount of information. This is when the *logical* idea of infinities hits a *physical* limit in nature. If objects—and their locations—are created by dividing through distinctions, then an infinite amount of information would also need an infinite space. That infinite information will take an infinite time to encode, even if it was physically possible to encode it. No mind could ever comprehend an infinitesimal meaning and no mind could create such meanings, unless we have an infinitely long lifespan. This gives us a connection between mind, matter and mathematics. We can define matter as the ability to encode information and mind as the source of that information. The limits of information generation in the mind and the limits of information encoding in matter are now also the limits of how small distinctions can become in mathematics. If the mind is limited by how small it can think and matter is limited by how small it can encode, then these limits are also those of mathematics.

Nearly two millennia ago, the Greek philosopher Zeno of Elea used a similar type of argument to discuss the problems in the idea of motion. Zeno's argument analyses the motion of a tortoise. Zeno claims that if a tortoise moves from X to Y, then before it reaches Y, it must move half-way to a point in between X and Y. But before the tortoise moves half-way, it must move one-fourth of the distance. And before it moves one-fourth the distance it must move one-eighth the distance and so on. Zeno concluded that there are infinite steps in moving from one point to another and the *computation* of these infinite steps will take an infinite time. In classical physics, we believe that the time taken to move one-fourth of the distance must be half of the time taken to move half of the distance. But from an informational standpoint, the amount of information needed to divide a length into four parts is twice the amount of information needed to divide the same length into two parts. While it appears that the object needs a lesser amount of time to go through a smaller distance, it takes more information to construct these smaller distances—provided we begin by assuming the reality of the whole distance. If, in fact, the tortoise computes the distance before moving, then he would be immobile because he could never compute the smallest possible distance. Indeed, we might say that the tortoise simply leaps from one point to another since that requires lesser information.

The size of the leap, however, would depend on the amount of information required to change one state to another. The state changes must be discrete, but they can be arbitrarily small or large, if the distance between the states is covered by the change. The problem here is simply that the computation of a smaller number takes more time than the computation of a larger number, if the numbers are being computed by *dividing* large into small. If therefore motion depends on the computation of infinitesimals, it makes motion impossible if each step of motion must be computed and there are infinite steps to be computed before anything moves.

While Zeno saw the impossibility of motion through this argument, he did not make the converse observation, namely that the non-existence of motion is also logically contradictory. Recall our earlier point about a moving machine that computes the denial of motion using the motion. If computation is motion and a statement denying motion could be generated, its very existence contradicts what it means. The resolution of this paradox would require us to concede that everything that can be computed may not necessarily be true. A statement denying motion can thus be computed even though the statement denies the method that led to its existence.

Zeno's paradox and other Greek questions about the origins of motion paralyzed Western scientific thought for nearly 2000 years until Newton postulated the first law of motion which claims that objects move at constant speeds unless accelerated by a force. Newton called this constant motion the property of momentum, and it became a cornerstone of the later development of classical physics. Newton's idea is remarkable because in Greek times philosophers were trying to figure out how an object moves to the immediately next position from its current position. And Greeks realized that there are infinitely many points between any two points because of which we cannot answer this question simply because we can't determine the very next position. How can an object move to the very next position when it cannot know what the next position is?

Newton's idea was to separate the motion of particles from their current location. He called these momentum and position. A particle always has a definite position, but the particle also has a momentum which defines the speed and direction of motion. With a momentum, the particle doesn't need to know the very next position. Newton reasoned that the very next position may be very hard to know, but it doesn't make a difference to motion itself if the particle just keeps moving. It will eventually pass all the positions, and we don't have to worry about how it goes to the immediate next point. This idea was later made rigorous using infinitesimals and limits by Karl Weierstrass, and that formalization is different from Newton's idea of motion. In Newton's theory, momentum is a different property than position. In the mathematical formalization, a trajectory is simply built out of points. If a trajectory is built from infinitesimal points, then it requires an infinite amount of information.

This idea is now known to be untenable in quantum theory where matter can occupy only discrete states, rather than continuous ones. That would imply that a particle does not move through all the states; rather it jumps from state to state. Newton's idea that postulates position and momentum as separate variables now seems more tenable than the reduction of motion to a succession of points. However, the idea in classical physics where position and momentum are separate variables does not exist in atomic theory because position and momentum are two *representations* of the *same* state, not two independent variables that together represent a state.

This new situation can be understood if a quantum is seen as a symbol. The position and momentum representations can now be understood as data and text interpretations of a bit. In physics, this is equivalent to saying that each object has two distinct features: (a) it has a spatial position that denotes the symbol's conceptual meaning, and (b) it has a momentum that denotes a program or the actions and the path by which it was created. The complementarity of position and momentum in quantum theory can now correspond to the notion that the same bit is both a description and an action. Additionally, descriptions and actions are defined semantically as meanings.

The difference between the ideas of momentum and a program is that a momentum causes an object *itself* to move while a program transforms *another* object. This implies—consistent with atomic theory—that an object by itself is stationary unless another object acts upon it. Newton's idea that if an object has momentum then it must be changing state is false in atomic theory because objects that have a definite momentum are still in *stationary* states. The conflict between the ideas of momentum and stationary states can be resolved through a semantic notion in which a quantum has the *potential* to change another object, but it cannot change itself. All individual objects are discrete, and this discreteness implies that the programs they represent are also discrete. No object can therefore cause an infinitesimal change to another object. Since objects cannot cause changes to themselves, there are no infinitesimal changes.

The reduction of a line to an infinite set of infinitely close points is false because infinite sets require an infinite amount of information and infinite time for construction. Infinity can be logically consistent only with the assumption that a line is built up from *a priori* real points. This idea is non-semantic because such points can never denote meanings (meanings are defined by their distinction from other meanings). To make the notion of a point semantic, it must be constructed by the application of distinctions. Each distinction represents a finite amount of information and an infinite amount of information cannot be encoded in finite space-time. The logical notion of continuity and infinitesimals in current mathematics is based on a physical and non-semantic notion of objects. When semantics is added to matter, the solution is inconsistent.

In that respect, the modern reconciliation of Zeno's paradox doesn't truly address the key problem arising from infinities. This is because Zeno's main argument was that if there are infinitely many points, then knowing (and thus computing) all those points will take an infinite amount of time. The use of infinitesimals doesn't address the infinity of points and the time it takes to compute the very next location, if in fact the next location is being computed.

Newton's resolution is different because in classical physics a particle has two independent properties—momentum and position. And momentum does not reduce to position; two properties suffice to describe motion. But even this solution is false in atomic theory where position and momentum are representations of the same state. Indeed, in atomic theory, a quantum has a momentum, but it is in a stationary state. Once we remove infinities in mathematics—since they are physically untenable—we can provide a semantic notion of change caused by programs. But this change will represent the process by which they were constructed, and the changes they can bring to other objects, rather than cause their own change.

Quantum theory has a counterpart in everyday language. This is where we associate two types of concepts with all things—object-concepts and action-concepts. We say that a knife cuts, an oven cooks, and a truck transports. Here, knife, oven and truck are object-concepts while cutting, cooking and transporting are action concepts. Note that many things can be used to cut, not just knives. Many things can cook, not just ovens. And many things can transport, not just trucks. Likewise, a knife, an oven and a truck may do other things besides cutting, cooking and transporting. The object and the action concepts are not identical although they are related. This insight is confirmed in quantum theory where position and momentum representations are complementary representations and denote the same state although described in two distinct ways. Thus, a knife can be used to cut, pierce, or spread something on a piece of bread. If we took one of those actions—e.g., cutting—then there will be many objects—like scissors, cleavers and blades—that could be used. Defining the form of the object—e.g., a knife—restricts the possible uses—e.g., cutting, piercing and spreading—so knowing its form does not fix what it can do, although it restricts them. Similarly, knowing a specific use does not fix the conceptual description of an object, although it restricts the possible descriptions.

In Zeno's paradox, motion is derived from positions. In Newton's solution, position and motion are two distinct ideas. In atomic theory, position and motion are neither equal nor totally unrelated. Both Zeno's paradox and Newton's solution are physical notions of the world without meaning, but the atomic notion is consistent with everyday semantic ideas about forms and their relation to action. The quantum solution shows that there are two distinct properties which are not identical and yet they are not entirely independent.

Zeno's paradox opens several interesting problems that have been debated for the last two millennia. There are at least three different and contradictory solutions to these problems that we know today—from mathematics, from classical physics, and from quantum theory. The quantum solution is identical to the everyday view of object and cause and their interrelation, but it requires a semantic interpretation of quantum theory, which I do in the book *Quantum Meaning*. The quantum problem also discards the notions about motion which led to Zeno's paradox and the idea of continuity which was seen, for a while, as a solution to the paradox. Without continuity and motion, *changes* must be seen in a radically new way. Now it is possible to see change as the evolution of meanings, which can represent the evolution of ideas or knowledge.

Counting Symbols vs. Counting Objects

Numbers are properties of collections. However, these collections are defined to be collections of objects rather than collections of symbols. Independent objects cannot denote names and concepts because each object only describes itself. The identity and the distinction of a symbol from other symbols depends on the other symbols within a collection. In a theory based on objects, the whole is a linear sum of its independent parts. But in a theory based on symbols, the parts are defined only in relation to the whole. This shift in thinking changes our view of numbers, because numbering is based on counting which depends on distinguishing and ordering. If the way a symbol is distinguished and ordered in a collection depends on the other members of that collection, then a symbol cannot be given a label or number independent of the other symbols in a collection.

When we count marbles—and for the moment let's suppose that marbles are independent of other marbles—any marble could be the first marble. However, when we count books, paintings, musical pieces or scientific theories, we order them by their meanings. For instance, books on mathematics might precede books on physics, which might precede books on chemistry, which are ordered ahead of books on biology, etc. Similarly, a scientific theory is understood by the changes it makes to current theories. We understand symbols not just by what they *are* but also through what they are *not*. In the objective-materialistic view of nature, an object is completely defined by what it *is* and there is no need to know what it is *not*. All classical physics was based on the idea of what an object is, which is defined independently of context. Classical physics was not concerned with what an object is not. For a symbol, however, both what an object *is* and what it is *not* must be defined, because the meaning of the symbol comes through its distinction from other symbols, and these distinctions are contextual.

Therefore, we cannot know what a symbol is not without the other symbols in the context. Since counting occurs within a collection, physical and semantic methods of counting present contrasting pictures about number theory.

Let's illustrate this with an example. Consider a family of husband and wife. These two individuals are defined by the presence of the other and married people undergo adjustment for the marriage to survive. People who want to avoid the pains of adjustment look for 'mutually compatible' partners. When a child is born or adopted into the family, more changes ensue, and the addition of the child not only changes the parents because they have a relation with the child, but it also changes the relation between parents. If a grandparent enters the family, it again alters the relationships. When children leave the home for higher education or when they get married, again relationships between members of the family are altered. More drastic changes generally follow if there is a divorce or death.

The point of this illustration is that we cannot describe members in a family as mutually independent marbles that can be added or removed without changing the other marbles. We must rather describe them as symbols whose meanings depend on the other symbols in a collection and include what a symbol is and what it is not. Through such relations, the members of a family acquire knowledge and intentions and become concepts and names. The identity, individuality and meaning of a symbol changes when the composition of a collection changes. The meaning of a symbol is different if there are two symbols within a collection versus when there are three, four or five other symbols within the collection.

A close mathematical analogue of this idea exists in the Hilbert Space theory of complex functions in which the space of possible functions is partitioned into an orthogonal basis of functions from which any other function can be formed through a linear combination. These functions are also called *eigenfunctions*. Hilbert Space theory is used in quantum physics to describe the behavior of atomic objects. I have separately interpreted the quantum formalism in my book *Quantum Meaning: A Semantic Interpretation of Quantum Theory* treating quantum theory as a theory about symbols rather than objects[17]. Classical physics is a theory about independent objects, but its description is inadequate for quantum physics. The Semantic Interpretation of Quantum theory describes that this is because we are dealing with symbols rather than objects. While objects are defined independently, symbols are only defined collectively.

The Hilbert Space formulation also shows that there isn't a unique basis of functions. Rather, an ensemble of particles can be described using many different eigenfunction bases. In each such basis, there can be a different number of particles whose meanings and identities will be different. The classical physical notion that an ensemble has a fixed number of *a priori* real particles, whose identities and total count are fixed, is therefore false. Given a collection, we cannot know the total number of particles in that collection and therefore we cannot order them into a sequence. The same collection can be partitioned into particles in many ways, and we cannot attach a cardinality or ordinality to the set (cardinality represents the total number of objects in a set, and the ordinality represents how these objects are ordered and counted).

A semantic theory of numbers must look like the Hilbert Space formulation of functions (which can denote objects). However, unlike the Hilbert Space formulation where an orthogonal basis of functions is created by the Hamiltonian—which represents energy and force—the theory of numbers must provide a logical rather than a physical theory. I believe that this logical notion can be meaning which is encoded as distinctions to create the distinguishability and individuality of objects. These objects are, however, not independent things as in the view of nature today. Rather, a symbol has two parts, which indicate what a thing is and what it is not. Both these parts are defined in relation to other parts in the collection. It is therefore incorrect to think that there are independent objects which can be collected. It is rather better to think that there is a collection which is *divided* into parts by applying informational distinctions.

This collection can be thought of as a more abstract concept, and the distinctions are divisions within that concept. It could also represent a context—such as a house—in which the rooms become the divisions although they are now designated by concepts such as bedroom, study room, bathroom, etc. The abstract concept is an undivided domain, and the distinctions on this domain are like the metric structure on that domain that produces a space. While this metric can be described as the distance *between* the space-time points (after the space has been created), the semantic view requires that the metric structure be treated as being logically prior to the points. When points—i.e. the states in space—are produced from the metric structure, then each object is *defined* in a manner related to the other objects. It follows that if there is no metric structure, then there are no individual states and hence no objects. Now, we cannot say that the objects 'move' in space-time. We must rather say that the information content—which divided space-time into objects—has changed. The change in information appears as motion. But it should be modeled as information change that results in motion.

Distinguishing in Space

When we count a collection of objects, we map some set of objects to the sequence of natural numbers, and after the mapping, we call the numbered objects as '1', '2', '3', '4', etc. But, before we label objects with natural numbers, we need to *distinguish* them. This is because unless an object is identified from another based on some criterion, there is no way to label them distinctly by numbers. In everyday life, this is achieved because distinct objects occupy different locations in space and time, and an object is considered distinct from another because of its location and/or time. Locations in space and time thus represent the foundation of counting things. By this counting, we name each object by its location. Therefore, space and time are the methods by which we *distinguish* things. Once they have been distinguished, they need to be *sequenced*—i.e. given numbers. This sequencing involves the selection of a coordinate system. Different coordinate systems can sequence the objects in different ways—i.e. give them different numbers (ahead or later in a sequence).

Now, we must remember that modern mathematics doesn't deal with object *distinguishing*; it only deals with the sequencing problem; i.e. giving different numbers to objects by assigning a different coordinate system; geometry is the study of these coordinate systems, not the study of 'space' which makes things *distinct*. It is our measure or description of the distinction, and it can be incomplete (because all the properties used in distinguishing objects may not be present as part of the space or time being used to count) or even incorrect (because the natural distinction between things may be incorrectly represented in the chosen coordinate system).

The difference between the choice of a coordinate system and the objectivity of a space in which things are fully distinguished has been obscured in modern science due to theories such as relativity because we started supposing that space and time were only coordinate systems that are used to sequence rather than domains in which objects must also be distinguished. For instance, in atomic theory, the *distinction* due to mass and charge is in the Hilbert space and time, which we cannot observe, while the *sequencing* of the particles due to observation is in the Euclidean space and time that can be measured. This separation between the spaces leads to the problem of probability in which we cannot predict which particle (in the Hilbert space and time) will arrive when and where (in the Euclidean space and time). Some scientists suppose that the Hilbert space and time represent the 'reality' while the Euclidean space and time represents the 'observation' and the quantum problem is that we cannot map the 'reality' into the 'observation' completely.

But I would diagnose the problem differently; it is an outcome of separating the spaces for distinguishing and sequencing. Sequencing is not a problem of observation. It is the problem of reality itself. Nature itself is a space and time in which objects are both distinct and sequenced, and our observation reveals that objective reality. The problem is due to our mathematical formulation of reality in which distinction and sequencing are separated, because we started out by supposing that space and time were not responsible for the distinction; they were only responsible for sequencing. To fix this issue, we need to reconceive space and time as domains that are responsible for both distinguishing and sequencing. For example, the fact that a particle has a different mass or charge must be described by a different location in space. Indeed, every property that we can observe or treat as being physically relevant must be embedded in the structure of space. To do that, we must reconceive space. The evolution of this reality then can adequately represent time.

This new idea about space and time is that of a tree, rather than a box. When space and time is a box, then arbitrary coordinate systems can be used to count, which makes ordering a choice. However, when space and time is a tree, then the root of the tree is a fixed origin; the trunks emanating from the root are the dimensions at the root; if the trunks emanate greater or lesser number of branches, then the dimensionality of space changes accordingly. The tree rests in a three dimensional 'box', as we observe, sequence, or count. But this box is inadequate for distinguishing because the box doesn't carry all the properties (e.g. mass and charge) of the objects. Furthermore, due to the choice of coordinate systems, the counting can also be arbitrary. On the other hand, in the tree we can both distinguish and count; indeed, the two are identical constructs because the distinction is defined in relation to the higher node in the tree, and the counting follows from the root to the leaf. So, the methods used for distinguishing and counting are identical.

The tree structure is more complete because it can be used for both distinguishing and counting. The box structure is incomplete because it can only be used for counting and that counting is also incomplete (because some properties like mass and charge are missing) and arbitrary (because the coordinate system is arbitrary). This is not to say that the world could not be looked up as a box; since the tree rests in a three-dimensional box, we could neglect the hierarchical relationships between the locations and (incorrectly) treat all locations as being equivalent. The result of this neglect would be that we could count incorrectly—i.e. put something ahead of the other. The total *number* of things will still come out correctly, but the *sequence* in which they should be counted would be incorrect.

This is pretty much where science stands today. In atomic theory, for example, we cannot predict the sequence among the particles because we don't treat the distinction between the particles hierarchically. We can still count the total number of particles but not the sequence in which they appear. The sequence is then described probabilistically which is equivalent to saying that we know how many types of particles exist, but not their interrelations.

The right approach to this problem is that we must treat the three-dimensional experienced space as a caricature of reality because it neglects the relations between the locations. A location in the real space is not directly related to every other location. That *structure* of relationships defines the distinction and sequencing. The multidimensionality of space can denote many observed properties (e.g. mass and charge). Since distinguishing comes before counting, there must be a domain in which things are distinct before they are counted. In fact, distinctions imply sequencing. Therefore, counting is an effect of distinguishing, but we must seek the cause. If the cause is not known, then the effects cannot be completely predicted.

The Tree Coordinate System

Let's imagine a scheme of counting objects using the spherical coordinates—i.e. $\{r, \vartheta, \varphi\}$. Since we are going to construct a tree, we are going to define these coordinates relatively—i.e. in relation to a conceptually more abstract node. Thus, the variable r represents the distance to such a node, the variable ϑ denotes the angle relative to this node in the horizontal plane, while the variable φ denotes the angle relative to this node vertically. If we disregard the vertical angle for the moment, in the horizontal plane, r represents the quantum of action, and ϑ denotes the *type* of the action. The quantum of action takes the new object a certain distance away from the more abstract node, but the angle of action ϑ produces a different direction away from the abstract node. The abstract concept represents the whole, and the variables r and ϑ represent the various *parts* of this whole. For example, if the abstract node was 'white', then the different parts of this concept could include 'red', 'blue', and 'green'. Note that we are in a horizontal plane with two variables—r and ϑ—so we can expand the concept of 'white' into its compositional concepts. Using the same scheme, we can again expand each of the components into further components and thus produce a tree of conceptual diversity.

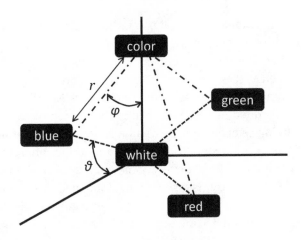

Figure-5 Constructing the Tree Coordinate System

Now imagine that we add the third variable φ into this space. The more abstract node would now represent the concept 'color', the variable r would be modified to represent the distance between 'color' and individual shades such as 'red', 'blue', and 'green', while the variable φ would denote the direction from 'color'. Thus, we can envision how a space of concepts could be constructed using coordinate systems. We must however remember that even though we have used a geometrical scheme to denote ideas, we are in a *semantic* space; which means that distances and angles must be treated as denoting meanings, rather than just quantities. It doesn't prevent us from visualizing them as quantities are representing them *as if* we were in a physical space, but these are *also* meanings. If we remained in the physical space, then the entire point of this scheme would be moot; we would still be within the mathematical paradoxes. By treating them as meanings, the location is identical to meanings, which allows us to interchange modes and overcome incompleteness and that must remain the singular purpose for this scheme.

The same type of scheme can also be used for *references*. But one additional consideration is required. When we relate to the world, the identities of both objects in the relation are modified. As an example, when two people relate to each other as parent and child, or as husband and wife, they are mutually defined by the relation. The relation gives them a new identity (of being a father, child, husband or wife). The relation is therefore not something *in addition* to the independent objects. Rather, that relation defines the identity of the objects themselves; in a sense, it is the object itself. Our references to the world therefore cannot be treated as pointers to things outside. They must rather be embedded in the object's identity. We can of course name the object being referred, but that name is defined in relation to the referrer. Thus, it is possible to give the same thing different names, provided we know the *context*—i.e. the referrer. For example, in a group of friends, you can call somebody 'John' because there is only one John within that group. The context unambiguously identifies the person, although it is not a universal identity, because there are potentially many people with the name 'John'.

The main point is that 'relation' or 'structure' is not something that exists *in between* things. It rather exists *within* those things, because each thing is defined by its relation to other things. We can say that each thing has a *form* due to its relationships. If those relations are altered, then the form of the object is altered.

The identity in relation to another object is different from the conceptual identity. For instance, conceptually, a tiger and a leopard belong to the cat family. But relationally they can be competitors. The identity of being part of the cat family and being a competitor are two separate kinds of identities, combined in the same individual. The conceptual identity is universal, but the relation is contextual. Thus, whether a tiger and a leopard are competing, they will remain part of the cat family. The competition is incidental to the context. This fact just indicates that we can distinguish and sequence things in more than one way. These methods of counting can exist simultaneously, but they are not identical; they remain complementary. This fact brings us to the earlier issue we saw in the Barber's paradox, which was that we were confusing the universal idea of a barber with the individual barber. The same name—or number—can be used in both types of schemes, although they are complementary.

The conceptual term 'barber' is only a potentiality; it represents the ability in the person to shave others, but that doesn't mean that the person is always shaving others. Similarly, there is potentiality in customers of being shaved; but they are not always being shaved. The universal attribute of being a barber is realized only in relation to customers when the barber is shaving the customer. However, the meaning of 'barber' remains universal—i.e. one who shaves others. In one sense, the truth associated with the meanings is universal. However, the truth associated with the individual is contextual. Since we can use the same name—or number—to describe both the universal and the contextual facts, there is a potential for confusion about whether the name refers to the universal or the contextual.

We can avoid this confusion by only dealing in the universals, but that means that we can only talk about concepts, not objects. If we cannot speak of objects, then we cannot form object *collections*, and we cannot use a collection to produce a conceptual definition. We must rather define a concept using other concepts, not using objects. This definition of a concept in relation to other concepts must not be circular, which is effectively saying that the definitions of concepts must form a tree structure (since a tree is free of circularity).

The other alternative is that we recognize that the universal 'barber' is simply a possibility, and it is converted into a reality within a context. When it is real, the universal and the contextual become identical, or the conceptual definition of shaving others is realized in a context when the person shaves others, and the appellate of being a barber (i.e. the relation to someone who is being shaved) can be legitimately applied. The universals are therefore pure possibility, and if we deal in these universals, mathematics is free of paradoxes. Similarly, the real world combines the universal and the individual, and if we study this real world, it is also paradox free because the names used to identify are identical to the universal concepts.

What we cannot do is treat the possibility as a reality in a context where it isn't. In that sense, the paradoxes are imaginary, and they arise when we make the mistake of defining concepts as collections of objects, because the object is always supposed to be real and that gives the false impression that something must either be a barber or not a barber, when the fact is that it can be both. For instance, someone can have the ability of acting like a barber and may therefore be called a barber. But he isn't currently acting like a barber, so he isn't a barber. The logical contradiction between someone simultaneously being a barber and not a barber is resolved by realizing that one of them is a possibility and the other is a reality. This distinction, followed by either restricting mathematics to universals (the possibilities), or applying it to reality (when the possibility is realized), frees us from mathematical paradoxes.

The Role of Contexts

Contexts represent when the possibility is converted into a reality. For instance, someone is a barber in fact when he is shaving others. Otherwise, he may have the ability of shaving others which is not realized. To realize this ability, there must be a customer who is being shaved. And this relation between the barber and the customer converts the possibility into a reality. In one sense, the ability of acting like a barber is perpetual (so long as the barber lives!) but it is observable only in some contextual relation to others. Therefore, the reality of being a barber is contextual, rather than universal.

Each individual can therefore be regarded *materially* as a collection of abilities, and each context as something that selects one or more of these abilities—perhaps one after another. We need two new ideas to relate to this viewpoint. First, we must treat matter as possibilities or abilities, which exist in a dormant form, and are utilized only in some relations to other such material objects. The 'object' in question is not *observable* because it is merely a possibility. It exists objectively, but it is not itself perceivable unless the ability is utilized in relation to something else. Second, to convert the ability into a reality, we must also treat the relations to other things objectively. These too are possibilities because each object can potentially have relations to many different objects, but those relations must be selected for an interaction and conversion of the ability into a reality. Therefore, what we call the 'universal' and the 'individual' are both possibilities, but their combination produces an observation. We call this observation 'reality' but it is not so. The reality is something that exists objectively prior to observation. In that sense, the ability and the relation are objective and real. Their combination produces the phenomena but since the same ability can produce many phenomena in different relations, it is an effect. The underlying causes—i.e. reality—are the abilities and relations.

Material properties become representations of meanings when they bear a certain relation to other objects. We have already spoken about relational meanings in the context of human relationships, such as father, mother, brother, husband, wife, son, daughter, etc. The body of the person in question is materially the same, but it acts differently in different contexts. This relational effect can be seen even in relations to inanimate objects. For instance, we might sit on a block of wood, and then it becomes a 'chair'; or we might keep books on that same block of wood, and then it is termed a 'table'.

Think of a red light at a street intersection. It is a material property. But in relation to us it represents the instruction 'stop'. Similarly, the green and yellow lights are material properties. But in relation to us, they represent the meanings 'go' and 'pause'. Now it seems that the name for an object cannot be entirely derived from the object's possessed properties—i.e. in this example, the frequencies of light. We must rather also consider their relations to us. We might note here that the colors 'green', 'yellow' and 'red' are also concepts or meanings. But they are different *kinds* of meanings than the meanings 'go', 'pause', and 'stop'. The former are material meanings while the latter are relational meanings. The color of the light itself cannot explain how people stop, pause, or go at a certain light color. Therefore, physics, as it is defined today would not explain why people behave in different ways at different light colors. We need to induct a new category—relations—to explain this behavior.

Choice in Nature

Once we recognize the fact that reality is produced from a combination of an ability and a relation, we can now ask the next important question: How do we select a relation? Given that there are potentially many possible relations, what determines which of these many possible relations are used to convert an ability into a reality? This is when we must invoke the idea of *choice*. But before I speak about choice, I would like to distinguish it from another common and confusing idea called *free will*. By its very definition, free will is supposed to be free; for instance, if the relation to things were decided by free will, then I could potentially interact with anything in the universe. Choice, on the other hand, is not free; it is rather constrained by two factors. First, there is a limitation on the number of things that we can interact with. During the waking experience, for instance, we typically interact with things to which we are physically close. While writing this paragraph, I'm tangentially aware of the sounds in the background, but I'm not focusing on them. That is, I haven't established a relation to them although I could. But I cannot obviously hear sounds that are being produced far from me. Therefore, there is something that limits the possible relations I can hold, and the choice is not entirely free. Second, I'm also limited in my choices due to my own nature; I'm interested in some things and disinterested in others. We seek happiness in life, so things that produce happiness are interesting to us. These interests may also evolve with time, and therefore the choices can evolve too.

Therefore, choices are not free will because they are constrained both by my interests and the availability of relations. Within these constraints there are still multiple alternatives (in general) that could produce different observations. Of course, in some situations, the intersection of my abilities, the available relations, and my interests may completely determine the observation, and it might seem that I have no choice. But, even in such situations, there is an important role played by my interests or desires; these are not fixed, although they may have been acquired in the past and may have become habits of choosing. Keeping aside the limitations of abilities and relations, we can strictly identify 'choice' with my interests and desires.

We are accustomed to calling these interests our 'nature'. Sometimes we also call this our 'personality'. It is objective in the sense that it exists—although in a potential form—like abilities and relations. To combine an ability with a relation, we need the third component of choice, personality, interest, or nature. This nature comprises many kinds of interests and although they persist in a potential form, they are not necessarily always active. Sometimes, the relation to something may excite our interest; sometimes our interest may drive us to choose a relationship; and sometimes our abilities may prompt us toward a relation or interest. In that sense, there isn't a strict causal relation between these three components. However, their combination is necessary to produce a phenomenon.

You can imagine a child playing with lego bricks. Each brick is a material object. However, there are many ways in which these bricks can be combined, which constitute different relations between the bricks. Some combination may produce a car, another one a house, and yet another one a tree. One of the many alternatives of such combinations are picked by our interest in creating something. Therefore, there are material components, relations, and choices, and their combination produces an object. In modern science, we presume that material components—i.e. the lego bricks—are the sole cause of complex material constructions. We forget the role of relations and choices, because we presume that 'forces' of nature are automatically responsible for some combinations over others.

This is a mistake. It is certainly true that you cannot combine arbitrary material components, just as you cannot glue lego bricks in an arbitrary manner—e.g. you cannot stick the bricks side by side unless you have other bricks on the top or bottom holding them together. So, material components *restrict* the combinations, but don't *exhaust* the possible relations. This fact is well-known to molecular biologists where large and complex molecules have many *conformal* configurations—i.e. the same number of atoms or lego bricks that can take many different forms just like lego bricks can form a house, tree, or a car. Once we recognize that matter can be rearranged in many forms, then both the form and the choice of one out of many potential forms becomes important. We can observe the form, but it is not a material *thing*—even though it is comprised of material things. Just like the house, car, or tree comprised of lego bricks is not identical to those bricks, even though it comprises the bricks. Our ability to reduce the house, car, or tree to the bricks doesn't mean much because the form has additional properties that the bricks did not have. Different conformal configurations of the same molecules, for instance, exhibit different properties.

In a very simple sense, we can distinguish between the material components and the arrangement of these components. The laws of modern physics don't determine these arrangements; they only prescribe that the total amount of matter and energy—i.e. the total number of bricks or the total amount of plastic that forms these bricks—remains conserved. Within that restriction, it is possible to create many different forms, and one such form requires a choice. Therefore, the *distribution* of matter is underdetermined by physical laws, because these laws only end up describing—(a) the compatibility between the parts, and (b) the conservation of the parts. They don't restrict the possible forms that can be produced by these parts—constrained by compatibility and conservation.

Matter redistribution requires no additional matter and energy, and in this sense the causality in redistribution is beyond current notions of causality that depend on the transfer of matter and energy. When matter is redistributed, there is indeed a transfer of matter and energy, but this redistribution cannot be predicted by any theory because the theory is consistent with all the distributions. Matter redistribution is a choice in all physical theories, not a causally determined fact. This entails that the physical theory is incomplete in explaining matter distribution, which needs relation and choice.

In modern physical theories, reality is equated with the *invariants* in nature. Furthermore, in classical physics even the parts of an ensemble are invariant (this corresponds to the idea that the lego bricks are immutable). However, in atomic theory, the lego bricks are not *invariant*; rather, the ensemble as a whole—which represents the total energy—is invariant. Therefore, it is possible to *distribute* the total energy in many possible ways (one such distribution is the ability to represent the quantum wavefunction using many different *bases*). You can no longer insist upon the existence of a certain fixed number of particles (i.e. lego bricks). You can only insist upon the total amount of plastic (energy in this case). This plastic (or energy) can in atomic theory be used to construct many kinds of lego bricks, followed by forms that combine these newly created bricks. Therefore, the problem isn't getting better; it is in fact getting worse as we progress further in science.

In fact, now, the whole (i.e. the total amount of plastic) must be treated as being more real than the parts (i.e. the lego bricks made of plastic). The problem is that you can never observe the total amount of plastic without creating some lego bricks, considering the possible forms, and then choosing a form. So, when you observe, one of the many distributions is found to be real, and it becomes compelling to think that this is indeed how matter existed prior to observation. But this is contrary to what the physical theories tell us—namely, that energy is fungible, that many forms are possible after energy has been divided, and that one form must be selected over others.

The Role of the Mind

The possibility of forms and their selection by choice gives mind an important role in nature. Even inanimate objects are not free of the consideration of the mind because we must explain the origin of form and how precisely one such form exists over the others. Current theories of nature exclude the mind and scientists studying the mind believe that the mind would be explained based on the same laws as matter. That the naturalist view of the mind is mistaken is not difficult to see, simply because minds are able to produce false ideas. To produce false ideas, there must be a role for choice, because a logically consistent system can never produce false ideas. Therefore, if minds are produced from matter—which is supposedly governed by consistent rules of logic and mathematics—then minds could not produce false ideas. The ability in the mind to produce false ideas itself indicates the fact that a logically consistent mathematical theory could not explain the behavior of the mind. Note that we are not worrying about which of the mental ideas are true. We are simply trying to explain *how* the mind produces false ideas. That question is more fundamental than the question of how truths are validated.

The existence of the mind therefore presents us with two important problems. First, we need to explain how the mind can focus on one thing—i.e. form one relation over others—when material objects supposedly have a causal relation with every object in the universe simultaneously. Physical forces are not *selective* about the objects they interact with, but the mind is selective: It will interact with one possibility at a time. Second, we need to explain how the mind can hold ideas symbolically as representations of meanings. A material object in the everyday sense is a symbol; for example, an ordinary chair is the symbol of the idea of a chair. That symbolism becomes possible due to the combination of the universal and the individual. The idea may exist as a possibility, which means that it may not always be realized unless the right relation is invoked. However, the idea is objectively existing. Material objects on the other hand are just *things* rather than *symbols*. That is, they are individuals without the universal. Therefore, unlike ordinary objects in which we employ two ways of counting—the universal and the individual—in the case of material objects of science we only use one. These problems entail many revisions to idea about matter.

First, we must recognize that matter exists as a possibility and that possibility is ideas or universals. Second, this universal becomes an individual symbol of the idea in relation to other symbols. That's when we begin to say that the symbol has a *meaning*; it's not that the meaning suddenly emerges from a meaningless reality; the meaning was always there as the universal. However, there are potentially many possible meanings in relation to different symbols. That selection of the meaning by a relation makes the meaning contextual even though it was a universal possibility to begin with. Third, there is choice in picking out a relation, selecting one of the many possible meanings. So, the symbol and its meaning in question are due to the combination of a universal potentiality, a relation that selects one out of the many potential universals, and the choice of a relation.

Falsities are outcomes of establishing inappropriate relations to things. For example, you could sit on an empty cardboard box thinking it were a chair, and the box could break, thereby falsifying your belief. The problem is not in the cardboard box or in your body. The problem lies in the relation you establish with it. As we have discussed, each object is the combination of certain abilities. But to believe in the existence of abilities that don't exist is false, and that falsity is demonstrated when we begin to use the object according to the false belief. This implies that even though the ability exists as a universal, its confirmation is contextual truth, because it is confirmed only in a certain type of relation. The objectivity of the ability isn't changed in the process, but our beliefs about those abilities are falsified. To hold such beliefs, we must possess a mental picture of the object that is different from the object itself. That mental picture is not true, although it exists. This drives a wedge between truth and existence; everything that exists is not necessarily true.

To allow for this wedge between truth and existence, we must invoke the distinction between the universal and the individual even in the case of the mind. The existence of the idea is the individual and it is true as existence. However, the idea itself—or the universal—is false when applied in relation to an object. To think that chairs can be used to sit upon is not false. The falsity is to think that some individual object is a chair. Again, this falsity is the outcome a relation between the symbol of a chair in our mind and the symbol of chair in the external world. The relation of truth is equivalence between these two symbols, and the relation of falsity is the non-equivalence of the two symbols. When truth and falsity are based on relations between symbols, then the classical divide between rationalism and empiricism collapses. Rationalism—e.g. mathematics—is the testing of claims by proofs; it involves symbols and demonstrating that one symbol (or set of symbols) is equivalent to another. Empiricism too involves the testing of beliefs (which are one set of symbols) against the reality (which is another set of symbols). The novelty here is simply that we are treating the material world *symbolically*.

The Axiom of Choice

Mathematicians abstract real-world considerations involved in a theory of numbers (i.e. the fact that numbering depends on counting which depends on the sequencing of objects objects) into an innocuous statement about *choice* in mathematics. This is called the Axiom of Choice (AC) and represents the idea that the mapping between and object and a number (the method by which objects are ordered into a sequence) is a matter of choice. This choice can represent the result of observation or physical measurement. After measurement, the measured values can be used to *name* the object. There is however a problem. The act of ordering involves relations between things, and only some of these relations are true, although all such relations can exist. Choices therefore can produce falsities if the relations are not carefully selected. If we restrict ourselves to only those choices that create empirical and rational truth, the choice is considerably curtailed, although not completely eliminated.

As we have already noted, the accessibility to relations restricts choices, as do our interests or desires. But within this restriction are further restrictions—e.g. the restriction of truth; not everything that is accessible to us, and which we are interested in, is necessarily true. For instance, a cardboard box may be available, and if I'm tired, I might desire to use it as a chair. That intersection of availability and desire does guarantee a choice, but it won't work because our mental picture about the cardboard box being sturdy is false.

Since the condition of truth is produced by relations between symbols, we can broaden our notion of restrictions on choice. For example, it is possible now to say that some relations constitute *righteous* actions, while others are unrighteous. Similarly, we can say that some relations create compatibility between our interests while others do not; we can call the former *good* and the latter *bad*. By such extensions to the judgment of relations, we can apply further restrictions on choice—namely, that choices must be true, right, and good. All these judgments are *normative* because it is not necessary for us to always hold true beliefs, do righteous actions, or form mutually satisfying relations. And they are normative because they are setting bounds on choices; that is, we are urging ourselves to make choices that are true, right, and good, but not forcing them. Indeed, if choices were forced, they wouldn't be choices.

The conditions of truth, right, and good can be objective, but our adoption of these restrictions upon choice is normative. There isn't any fundamental contradiction between objective and normative, as has been supposed by many philosophers in the past; they argued that ethical considerations must remain outside science because they deal with norms, while science deals with the objective truth. The problem lies in their conception of truth as something that *exists* which rests upon eliminating meanings from the world. Once you remove meanings, then you must also reject *judgments*. You now restrict yourself to a limited idea about truth—namely, if something exists, then the claim of its existence is the truth. Commonsense shows that hundreds of false ideas exist in our mind, but their existence doesn't make them true. So, this false notion about truth as something objective—rather than something relational—underlies the rejection of right and good as valid judgments beyond truth. If we now correct our notion of truth—i.e. make it relational—then it opens us to even the judgments of right and good. This is progress because we are (a) admitting relations and choice, and (b) subjecting these relations and choices to various types of judgments.

The Axiom of Choice can now be modified to say this: Yes, there are choices which help us create form and order, and sequencing objects by counting them is the prototype of all order; however, all order is not necessarily true, right, or good. Therefore, the claims about choice do not end by supposing its existence. We must rather add *responsibility* to *choice*. The responsibility is that we must prefer truth, right, and good, over falsity, unrighteous, and bad. This is a normative demand and not a logical demand. That is, we cannot contradict choice by forcing truth, right, and good; but we can exhort the correct use of choice by attaching responsibility to it.

My point is that mathematics is overly restrictive by adopting an objective notion of truth, rather than a relational view of this truth. This objective notion works only by eliminating meanings, which then leads to logical paradoxes—such as the Gödel's incompleteness theorem—because we cannot in fact give up ordinary language and its categories such as individual and universal. Now, if we induct these categories back into mathematics, we must recognize that what we call 'objects' are in fact *symbols* in which meanings exist as potentials and they are converted into observation in relation to other symbols. The truth of these symbols is therefore not merely their *existence* but rather contingent upon the relation to other symbols. We must also recognize that these relations are subject to choices, because they too are possibilities waiting to be realized. Now, we have come full circle, and admitted—(a) the existence of meaning, (b) the existence of choices, and (c) a relational rather than objective notion of truth. It now becomes easier to extend this idea to other judgments such as right and good, without precluding choices. Judgments now become normative; and the novelty is that even truth is a normative judgment. That is, nobody is stopping us from choosing falsities, and false ideas can exist in our head. We are just being exhorted toward the choice of truth, right, and good.

I don't have a good name for this viewpoint, but I can assert that this approach makes mathematics complete in many ways. First, it resolves the paradoxes arising out of using meaning and naming, naming and activity, etc. Second, it takes us beyond the question of truth, into the questions of right and good, which are human questions, although have been left outside mathematics. Third, it brings mathematics closer to reality—namely by realizing that beliefs about truth cannot be compelled, and nature allows the existence of false beliefs, which would create contradictions in the study of the mind if we stuck to our existence view of truth; by advocating a normative notion of truth, we solve the logical paradoxes which must arise in a logical study of the mind. While the first of these considerations is within the scope of present-day mathematics, the other two are broadening its horizons, bringing it closer to reality in the sense that we understand it humanly.

4

Numbers and Meanings

The laws of Nature are but the mathematical thoughts of
God.

—*Euclid*

Natural Numbers

Some definitions of natural numbers include all positive integers beginning with 1 (1, 2, 3, 4, etc.). Other definitions include all positive integers beginning with 0 (0, 1, 2, 3, 4, etc.). The reasons for including 0 in the list of natural numbers have to do with the simplification achieved in Boolean operation and definitions of numbers.

For example, it is convenient to think of Boolean logical states of False and True as 0 and 1, respectively. However, the Boolean Truth tables are written in a way that produces some counterintuitive results as far as number additions are concerned. For example, in Boolean Logic, 1 + 1 = 1, which is patently wrong if we were to treat these as numbers. The convenience of Boolean Logic is that if numbers are represented in terms of Boolean numbers, then addition turns out to be the logical OR operation and multiplication becomes the logical AND operation. This is very convenient for computers and despite some counterintuitive results such as 1 + 1 = 1, the usage has stuck. Theoretical mathematicians too prefer including 0 in the list of natural numbers, because then it is possible to denote 0 by an empty set {}. The number 1 can then be represented by a set of one member—namely the empty set—e.g. {{}}. The number 2 by the set of two members— {{}, {{}}}, and so forth. But there is a serious question regarding what the empty set is. Can you measure nothing? You can certainly measure the presence of something. But can you measure the absence of everything? How would that be any different from not measuring at all, or measuring wrongly and not finding anything, because you did the measurement incorrectly?

Considering these conceptual difficulties, I prefer the definition of natural numbers that begins with 1. It represents what you can count by sequencing and the total number of objects that result from counting. It is convenient because it provides a 1-1 mapping between cardinal and ordinal numbers. For example, if you began counting from 0, and there were 5 things to count, then your ordinal counting will end at 4 while you will conclude that there were 5 objects. This problem doesn't arise when you begin counting from 1.

I want to warn the reader that I will revise this definition of number—as indicators of a *quantity*—a few times as we induct new types of numbers in subsequent sections. But, for now, we can begin by supposing that number means quantity, and since you cannot measure emptiness, therefore, 0 is not a quantity. Hence, natural numbers must begin from 1, and denote quantity measurement.

Once we set the lower bound on the natural numbers, we can look at the upper bound. Normally, we all suppose that there isn't an upper bound, or the upper bound is *infinity*—∞—which is supposed to be this unimaginably large quantity that you can never define or measure. If the problem of measurement was impeding us from accepting 0 as a natural number, then by the same measure, we must have the same problem in accepting ∞ as a quantity. Normally, and almost imperceptibly, we attach the idea of a *unit* to numbers. For instance, when we say that something is 100 miles away, we suppose that it is 100 times farther than something that is 1 mile away. This naturally leads us to the notion that ∞ is infinitely larger than 1. But suppose for the moment that we have a collection of objects which can be ordered from 1 to ∞ but each object is successively half in size than the previous one. As far as counting is concerned, we would still sequence them from 1 to ∞ and then say that there are infinite objects. But as far as quantity is concerned, the sequence of $1 + 1/2 + 1/4 + 1/8 + 1/6 + ...$ would add up to 2. This raises the question: should we regard number as quantity or simply *cardinality*?

Cardinality is defined as the number of objects in a set, and it can be infinite. But even a collection of infinite objects can represent a finite *quantity*. The collection of infinite objects leads to an infinite quantity when we assume that each object has the same *unit*. For example, our unit may be a mile or a foot, and if all objects have the equal size, then the total quantity is indeed infinite. But what does this unit have to do with number? It represents a pragmatic tool used in measurement of distance, weight, speed, etc. but in doing so, we give it a *dimension* such as distance, weight, speed, etc. The unit we attach to the number is due to this dimension, not of the number itself. If we strip the dimensions from the numbers, then we are simply left with the ordinal (first, second, third, etc.) and the cardinal (one, two, three, etc.) which have nothing to do with quantity.

This brings us to the first modification to our definition of a natural number: a natural number is not a quantity; it can be either an ordinal or a cardinal, and if we begin counting from 1, these two are identical. That is, we can say that we have many objects or a few objects. But they are not necessarily big or small in quantity. This is a subtle difference, but it helps us separate the idea of number from that of quantity. The idea was first illustrated by Zeno's Paradox[18] in which, as mentioned in the previous chapter, he argued the impossibility of motion by supposing that before a runner can go from point A to B, he must go half-way through. But before he can go half-way through, he must go one-fourth of the way, and before that he must go one-eighth of the way, and so on. This naturally leads us to infinite steps from A to B, if were to count each step as an individual thing in a sequence of things. And yet, these infinite things are present within the smallest thing. It is just that the *unit* of each such thing is successively halved, such that the total quantity is unchanged, but the cardinality and ordinality are infinite.

The revision is that a natural number is not a *quantity*; it is rather the act of *counting*. The sequence of counting is limited by how far we can go on counting—which then depends on the time at hand— if counting each thing takes the same amount of time (although that may just be our assumption). Effectively, that attaches a unit—of time—to the act of counting and thereby the counting till ∞ would then require an infinite amount of time (as Zeno concluded, arguing for the impossibility of motion). But if we are strictly dealing with numbers, we can stop worrying about this problem and just focus on the act of counting—i.e. sequencing from 1 to ∞.

Is Zero a Number?

In so far as numbers are used to count objects, zero is not a number. We cannot measure the 'size' of zero because that measurement result would not be distinguishable from not measuring anything at all. We also cannot order an empty set, because ordering requires the ability to distinguish and emptiness cannot be distinguished. The number zero therefore violates intuitions about cardinal and ordinal numbers. And yet, zero plays an important role in mathematics as the *additive inverse*[19] and *additive identity* in various types of algebras. In the simplest case of integers, the additive inverse of +1 is -1. If the algebra supports the operation of addition, then the number zero must be present in the set of integers described by the algebra. This is because adding +1 and -1 produces 0. If zero is not added as a valid integer, then there could not be negative integers because the following types of algebraic operations would be impossible:

$$-1 + 1 = 0$$
$$0 - 1 = -1$$

The existence of zero is therefore closely tied to the existence of negative numbers. However, negative numbers don't intuitively lend themselves to the idea of cardinality of a set. How can we measure the size of a set as a negative number? Negative numbers are also problematic as far as ordinals are concerned because in the increasing order of successors, the counting must begin at $-\infty$ and end at $+\infty$. How can we begin counting from $-\infty$ if we don't know what it is? It is practically therefore very difficult to attach an ordinal 0 to an infinite set, because we cannot identify which member of a set is right in the 'middle' of the order. Given these challenges in treating negative integers as cardinals and ordinals, we are led to ask: Do negative numbers really exist? This problem may not arise for the Platonist who views numbers as abstract objects, not real ones. But the problem is very tangible for the mathematical realist.

Historically, negative numbers became important when people were dealing with money. You could borrow money, and then you had a *debt*. How do you represent money that is *owed*? Accountants invented the use of negative numbers to represent debt. But factually debt is a *moral* notion: you *expect* that someone will pay the debt in the future, but now the money is physically with the person who hasn't yet paid it. Therefore, we are not talking about the money itself, but the *direction* in which it is going to *flow* in the future.

One way to generalize this idea is to always associate a number with a direction. For example, in geometry we can speak of a straight line with a zero in the center and speak about positive and negative directions. Negative numbers are now understood by the fact that a number is not just an order but also a direction, akin to vectors. If this idea is extended to a multidimensional space, direction itself becomes a number. For example, in a two-dimensional space, you can have two numbers—a quantity and a direction. This indeed forms the basis of some well-known coordinate systems of counting[20].

We are now led to a compelling picture in which a negative number is in fact two things—an order and a direction. Similarly, even a positive number must be defined by an order and a direction. However, if we look closely at the problem, we have only shifted the burden from natural numbers to direction because the natural numbers still begin with 1, while the direction can be plus or minus. But this is where the eminence of 0 comes into play. We are no longer counting from $-\infty$ to $+\infty$ as if there were a monotonically increasing order (as we commonly suppose with negative numbers today). We rather have two kinds of orders—from 0 to $+\infty$ and from 0 to $-\infty$. The idea of a direction makes sense only when we regard 0 as the origin from where we count. However, since 0 is part of both the positive and negative sequences, it is in a special position of its own.

This special position can be recognized if we see 0 as the sum of all positive and negative integers. By this 'sum' I don't mean a *quantity* anymore, since such quantities involve units, and we already rejected the attachment of units to numbers. I rather mean that 0 represents a cardinality although it is the cardinality of nothingness. Now, as we have discussed previously, the measurement of nothing is not a measurement. We cannot distinguish it from not measuring or measuring incorrectly and not finding anything. Therefore, the truth of the claim that nothing exists is indistinguishable from the falsity arising out of misperception or incorrect procedure. We should rather take a more nuanced metaphysical position in which nothingness is not non-existence. It is rather the *combination* of all existents—i.e. the addition of all the cardinals—such that we cannot distinguish them. Thus, zero is not *nothing* but *everything*. Since the integers have their opposites, if we combine them, they cancel each other out, which produces the appearance of nothingness.

Negative Numbers

A new problem now arises, which, as we will see shortly, requires us to change our definition of number yet again. This problem concerns the cardinality of negative numbers. If we cannot measure nothing, then we certainly cannot measure the absence of a certain number of things. For instance, we cannot say that I observe the absence of 5 things, because absence itself cannot be measured. It follows that I cannot treat the negative numbers as cardinalities. And yet, we can treat positive numbers as cardinalities. The solution to this paradox requires us to make a distinction between concepts and objects, or universals and individuals. It is possible to conceive of negative numbers as pure concepts, although we cannot treat them as the cardinalities of a set of things. Positive numbers on the other hand can be treated both as pure concepts and cardinalities of sets.

The solution to this paradox is that positive numbers are useful in counting objects—you cannot count a negative number of objects. However, negative numbers are useful only as concepts. If we were to provide a definition of number, it would just be a concept, or an ordinal, not a cardinal. Cardinality is simply a limited view of these numbers when it is applied in counting things or measuring quantities (by attaching units to them). This real-world application of numbers is too restrictive if we induct negative numbers. To regard negative numbers as 'real', we must forego the notion of cardinality. Cardinality also doesn't work for negative numbers consistently. For instance, a collection of 5 negative numbers has the cardinality of +5 rather than -5. The unique mapping between ordinals and cardinals which works so well for positive numbers fails miserably for negative numbers. This, however, only means that number is only a concept, or a universal. When this conceptual world advents in the real world, only a subset of this conceptual world—the world of positive numbers—appears. So, positive numbers can be used for objects, and they can be used for counting and measuring. But while dealing with negative numbers, we must adopt a purely conceptual approach.

This entails a minor modification to the idea of zero. As we noted earlier, we could treat zero as a cardinal, if it were regarded as the sum of all numbers—in short, a cardinal. But on closer look, -5 is not a cardinal; the cardinality represented by -5 is 5. So, in what sense are we adding +5 and -5 to produce a zero? The short answer is that the idea of cardinality applied to zero is not entirely sound. We need to redefine zero as a concept—the concept of *everything*—that produces opposites. Once these opposites are combined, we are left with nothing, and hence we call this a zero. But this combination is conceptual rather than a quantity or a cardinality. For instance, if we combine the opposite ideas of 'hot' and 'cold' the two will cancel each other and produce a state that is *neither* 'hot' *nor* 'cold'. We are accustomed to thinking in terms of one side of the opposite, but not used to thinking in terms of a category that is *neither* of them. Zero is that state which is none of the opposites; it violates the logical principle of mutual-exclusion in which a choice must choose one side of the opposite alternatives, and both alternatives can't be rejected. Through this violation of a logical principle, zero is outside logic, or the precursor to logic, which exists only when opposites do.

The rejection of cardinality has no impact on the ordinals. Since we already added direction to an order, we can say that we have two directions in which we are counting from 0. Therefore, it is perfectly meaningful to say that relative to 0, I am at the 5th object in the negative direction, although it is meaningless to say that I have -5 objects in a set. This is also how negative numbers are used in physical theories. For instance, we never say that we have -1 electrons. We rather say that we have one electron with a -1 charge, where charge must be treated as a method of sequencing things in an order—i.e. giving them locations in a conceptual space. The use of negative numbers as quantities must be rejected. So, for instance, we cannot say that the electron has -1 charge, presuming that this were indeed a quantity. We should rather say that the electron has the *opposite type* of charge. The revision in thinking is that we should stop thinking that the electron is *missing* some charge. We must rather think that there are two types of charges—positive and negative. In short, charges are ordinals and not cardinals.

The cardinal way of thinking implies that charge is only of one type—namely positive—and the particles with negative charge are missing that positive charge. The ordinal way of thinking tells us that there are two kinds of charges, and the particle with a negative charge possesses a different *type* of charge than the particle with the positive charge. So, instead of one type being present or absent, there are two types which can be present. The presence of one type automatically implies the absence of the other, therefore, the novelty is that there are two distinct types rather than just one type, and these are mutually opposed as *types* rather than quantities.

This way of looking at the world liberates us from the problem of negative cardinalities (we had already rejected the reality of negative quantities). But it also pushes us into a conceptual world rather than a physical world. In the physical world, we can only use positive numbers, but in the conceptual world we can speak of types or concepts which can be mutually opposed. Even if the concept of the negative type were converted into an object, it would still be an object to be counted using the positive number rather than the negative number, although it will possess *opposite* properties. In short, objects must be positive integers, but properties can be both positive or negative, and these properties are *ordinals*. Another way to state the same idea is to say that properties are conceptual rather than physical, whereas the objects or individuals are physical.

This view requires us to distinguish individuals from universals. We make these distinctions in physics today when we distinguish between a particle and its charge. The particles must be counted using positive numbers, but the charges can be positive or negative. When negative numbers are conceptual, then the particle with a negative charge combines an individual and a universal. There is only one *instance* of the universal, but there can be many instances of the individuals. Thus, for instance, negative charge is a universal concept, but there can be infinite individual particles with a negative charge. Charge is no longer a physical entity, but the particle is physical. This overturns centuries of thinking in which the particle was a pure concept while charge was a physical property. Factually, science never depended on the physicality of the particle; it only relied on the properties because natural laws are based on these particles. Now, if we overturn this thinking and claim that charges are conceptual (because they can be positive or negative) then the laws dealing with such properties must also be conceptual rather than physical.

This may sound odd as first sight, but it is not. It brings us closer to understanding why mathematics as a whole—which is supposedly conceptual—applies so well to the material world. How can properties be physical but governed by conceptual laws? How do the physical and the conceptual interact—which goes back to the mind-body problem—that has remained unresolved for centuries. This problem gets quickly addressed because both laws and properties are conceptual rather than physical. Now it is possible to treat physical properties (e.g. charge) as much conceptually as we treat formulae and mathematical laws conceptually. The conclusion, however, would inevitably be that causality is conceptual. That is, the cause of change is not driven by physical properties but due to concepts. The world we see and touch, taste and smell, is not physical 'stuff'—at least the *reason* we perceive it is conceptual. However, the reason we find individual things is still physical, although that physicality is not responsible for our perception or causality[21].

Therefore, the fact that there is an individual electron, or an individual table, is due to physicality, but the fact that it is an electron, or a table, is conceptual. Conceptually, there is only one electron or table—because these are pure ideas. However, physically, there can be infinitely many *individual instances* of these concepts. We can classify and categories the individual instances using concepts—thereby making science possible—precisely because they are comprising concepts. However, the concept in our mind and the concept in the external object—if our understanding is correct—is the same conceptual entity. It just has two separate instantiations—one within our mind and the other in the external object. As a result, there is no mind-body problem to worry about because both my mind and the external objects are conceptual *and* physical. The concept is the universal, but the individuality is physical.

This poses a problem of how the conceptual and the individual combine, but this problem can be easily resolved by postulating two kinds of spaces—one conceptual and the other physical. The location in the conceptual space denotes the differences between concepts, while the locations in the physical space represents the individuals. The conceptual space—just as in Platonism—is unchanging; it has all the concepts that we can imagine, which remain pure possibilities. When the location in the individual space is *mapped* to a location in the individual space, an individual object with properties is created. Without this mapping, the location in the individual space too remains a possibility, which means the individual exists, but it is unable to act without the combination with the universal.

By separating the individual and conceptual spaces, and then conceiving a mapping between them, we solve many problems:

- We resolve the issue of negative numbers being ordinals but not cardinals. We resolve why we always count individuals using positive numbers and never using negative numbers.

- We can explain why conceptual laws are useful in explaining the material world—the reason being that these laws deal with properties which are conceptual. They don't deal with individuals. Both the law and the property are conceptual.

- We can explain how the mind cognizes the world in terms of concepts, which would be impossible with a physical world. If the world is physical and the mind has ideas, then we are left with the intractable mind-body problem, which then leads to a subsequent problem about how true knowledge is possible. Given that concepts are universals, how can the *same* concept appear both in our mind and in the external world? If they are not the same concept, then true knowledge would be impossible. By saying that laws and properties are conceptual, it becomes possible to see why we know the external world correctly even though my mind and the external world are distinct.

- It gives us a tangible mathematical conception about how the conceptual and the physical interact; this interaction is a mapping between two spaces, which *instantiates* the concept. This is an important problem because mathematicians like to believe that their study deals with the world of possibilities rather than with the physical world. If a mathematical theory has no physical applications today, the theory is still valid. Indeed, many mathematical theories have been found useful *after* they were discovered (Hilbert Space in atomic theory and Riemannian Geometry in relativity theory are prominent examples). So, the fact that concepts can precede the physical application of these concepts to the real world should be considered a problem.

- Since some concepts may not be mapped to the physical world right now, we may not observe them—think of dinosaurs that we cannot see right now. Nevertheless, they remain a possibility and must therefore exist conceptually. Therefore, a dinosaur cannot be 'created' or 'evolved', but 'instantiated' from a universal pre-existing idea if it combines with an individual[22]. Just because we don't see a dinosaur doesn't mean it has ceased to exist conceptually; we can certainly *think* about it in our minds. It must exist as a conceptual possibility for us to think about it.

The main point that I would like to leave the reader with is that when talking about positive and negative numbers, we are dealing with two spaces—conceptual and physical. The physical space only involves positive numbers while the conceptual space comprises both negative and positive numbers. The need for zero arises only in the conceptual space. Alternatively, we can say that the physical space is the space of *natural numbers* (beginning with 1). But the conceptual space comprises positive and negative integers.

Complex Numbers

The existence of negative numbers becomes murkier if we realize that positive numbers have a square root which is also a positive number, but the square root of a negative number is not a negative number. To solve this problem, mathematicians had to postulate imaginary numbers, which are square roots of negative numbers. Complex numbers create interpretive difficulties because they have two parts—real and imaginary—which cannot be added in the conventional sense, even though we write them using the traditional sign of addition as a + bi. Of course, decimals also have two parts—the integer part and the fractional part—but unlike decimals which can be *visualized* in a single dimension (for instance 3 + 0.5 = 3.5)—complex numbers require two dimensions to visualize. The geometric visualization of complex numbers is shown in Figure-6. This picture sometimes leads us to the idea that real numbers are single dimensional while complex numbers are two dimensional.

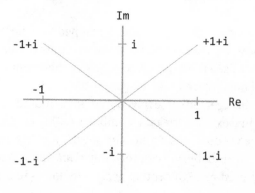

Figure-6 Complex Number Visualization

We should not be surprised by this representation because as we already saw in the case of negative numbers, a number is an ordinal along with a direction—e.g. there are two directions of counting from 0. Once we have added two directions, it is conceptually easier to add even more directions, because at least the idea that a number has a direction is no longer problematic! Positive and negative numbers just support two directions, but imaginary numbers add two more directions to create *four* directions on a two-dimensional plane. The combination of real and imaginary numbers (called complex numbers) creates infinite directions. Unlike the multiplication of positive numbers which only changes cardinality, and unlike the multiplication by negative numbers which flips the direction by 180^0, imaginary multiplication changes direction by 90^0 at a time. The whole thing would sound much less problematic if we realized that numbers must always have a direction.

A much more serious problem arises when we expand the definition of a complex number to multidimensional spaces. In a three-dimensional space, for instance, complex numbers form quaternions whose multiplications are written as follows:

$$i^2 = j^2 = k^2 = -1$$

$$ij = -ji = k$$

$$jk = -kj = i$$

$$ki = -ik = j$$

$$kji = ikj = jik = 1$$

Now, we are breaking the commonsense intuition that there are 3 dimensions of space and requiring that we have a 6-dimensional space (three dimensions of real and imaginary directions each). Why would such a numbering scheme ever be useful in the real world if the real world only has three dimensions? This problem can be solved by thinking of a *hierarchical* space in which the imaginary space is factually another space *outside* the real space, and the real number space is an *object* in that space. This object acquires a location and direction in the higher space, which means we are changing the direction and size of one space inside another space while doing complex number operations. We are no longer dealing with objects; we are dealing with objects that are also *spaces*.

With three dimensions, it is now possible to say that there is a three-dimensional space embedded in a three-dimensional space, although one of these spaces appears as an *object* inside another space. This hierarchy can be potentially infinite, if we expand the number of imaginary dimensions. For instance, we could have six imaginary directions and only three real dimensions. That would imply that an imaginary space is an object inside another imaginary space which then contains a real number space. As these dimensions increase, only the last embedded space is real; everything else is imaginary. They are however imaginary in a different sense each time because they contain all the previous embedded spaces.

The term 'imaginary' can now be given an intuitive definition—it is outside the current space, as it is a 'higher' space. The 'lower' space is contained inside the 'higher' space as an object. The hierarchy of spaces is potentially infinite, until we come to the end of the hierarchy. This entails the need for potentially infinite spaces, which would make the mathematics infinitely complicated.

Theories such as quantum mechanics—which utilize the complex number theory—restrict the problem to a single imaginary space but suppose that there are *infinite* objects in six dimensions. The right way to think about the problem is that there are infinite objects *and* infinite spaces. The same thing is a space containing the lower object, and an object contained in the higher space.

Since the same thing is both an object and a space—although in different higher and lower contexts—unless we capture the context, simply using 3 dimensions for object and 3 dimensions for space isn't going to work. This is the mathematical reason why theories such as quantum mechanics become incomplete. They are right in thinking that the quantum object must be described by complex numbers because that object is also a space, but the relation between the objects/spaces is missing. A hierarchical geometry is needed to solve this problem in which quantum objects are not in a single space, but successively contained inside a cascading hierarchy of spaces.

This idea is intuitively evident from the quaternion equations where the three dimensions in space can be produced from the unit scalar—i.e. an object in real space. Since the three dimensions represent mutually orthogonal dimensions, the equation suggests that something that exists in the real space (a number) can produce a space comprising three imaginary dimensions—*i, j,* and *k*. Thus, every point in the real space can be treated as the origin of another space, and since that point produces another space with locations within it, each of those locations can then be treated as the origin of another space. Space—if looked upon in this way—is infinitely enfolded (assuming the hierarchy of enfolding is infinite).

An intuitive way to think about this hierarchy is to think of the color white which can then be seen as the origin of three orthogonal types of colors—e.g., red, green and blue—each of which can act as a dimension. The color white is not in this space and is 'imaginary' from the standpoint of this 'real space'. Furthermore, if the location in this color space are not infinitely small *points*, then each location in this color space can again be treated as a space which then expands into three further dimensions, and if the locations in the newly created space are not infinitesimal points, then the process continues indefinitely. You can also imagine a similar type of process when thinking about how the world is divided into countries, which are then divided into states, that comprise cities, streets, and houses. The novelty here is that we don't divide the space into infinitesimal points; rather countries, states, cities, streets, and houses are 'locations' in an intuitive sense, which can be further subdivided. You need to emerge out of the house to be in the 'street space', emerge out of the street to see the city, and so forth, which is akin to 'rising' in the hierarchy of spaces. Imaginary numbers represent this intuitive way of thinking about space hierarchically with one difference—they are still treating locations in space as infinitesimal points.

In short, you try to collapse the hierarchy of spaces and produce one three-dimensional space, but the reason it doesn't work can be attributed to the fact that locations are not infinitesimal points. Rather, it is possible to treat a country as a 'location', even though it comprises infinite infinitesimal points. Imaginary number arithmetic can be real *if* the idea that space is infinitesimal points is false. This doesn't mean that we cannot divide a 'location' into smaller locations. But it does mean that macroscopic locations can exist in addition to infinitesimal ones. These macroscopic locations will then *contain* the infinitesimal locations. In short, we can treat a collection of locations as a location—consistent with the idea that a collection of objects can be an object. We are simply recognizing the reality of macroscopic objects in addition to those of infinitesimal ones, a move that puts us in contradiction with *reductionism* (in which everything must be reduced to the smallest conceivable parts) but that shouldn't be a problem, unless the idea of treating a collection of parts as a part of an even larger collection is itself a problem. When we recognize a collection of things as a thing, which is a part of an even larger thing, then we subdivide space into successively smaller parts. The larger space is 'outside' the smaller space and you cannot refer to it in terms of the same dimensions as you do the parts within the space.

Every location in space is both an *object* in a higher space, and a *space* from the perspective of the objects within that space. The ability to treat the same thing as a space (i.e. real) and an object (i.e. imaginary) is possible with the use of complex numbers, because we are adopting an anti-reductionist stance in which the whole doesn't reduce to the parts; it is rather an object in a higher space which cannot be described using the dimensions of the lower space.

Rational Numbers

While many rational numbers have infinitely long fractional parts, a rational number with infinite digits represents a *loop* because the digits in fractional part of the number repeat. Below are examples:

$$1/30 = 0.033333333333 \ldots$$
$$9/11 = 0.818181818181 \ldots$$
$$22/7 = 3.142857142857 \ldots$$

The fact that rational numbers can be represented by repeating sequences of digits means that they can be represented by *stored* programs (recall that stored programs violate semantics). The number 1/3, for instance, can be represented by a finite stored program. All rational numbers can be constructed geometrically using a finite number of steps. For instance, the number 22/7 can be created by dividing a line-segment into seven parts and then adding 15 such parts to the original 7 parts. Alternately, you can divide 1 into 7 parts and then add 21 of the same parts to the original part.

Georg Cantor proved a remarkable result[23] showing that rational numbers are *countable*—i.e. can be mapped to integers 1-1. Therefore, there were as many rational numbers as there are integers (assuming of course that the integers are infinite). We will see in the next section that this is not true for irrational numbers, so the result is significant because even the rational numbers can have infinite digits in them and of course the irrational numbers always have infinite digits in them. The important thing, however, in Cantor's proof is that he doesn't use the fact that rational numbers may have infinite digits; he just represents them as *fractions* which can be represented finitely (e.g. as 1/3) rather than as 0.3333... etc.

Rational numbers seem to reinvoke the idea of quantity, which we rejected earlier when dealing with negative numbers, because it assumes the existence of a *unit* of measurement, and numbers are not about measurement. We had also rejected the notion of cardinality when dealing with negative numbers because the cardinality of a set of numbers from 0 to -∞ is +∞, and this breaks the 1-1 relation between ordinals and cardinals that is true for positive integers. Let's stick to our prior claims and within those constraints find an original way to understand the rational numbers as pure ordinals.

In describing complex numbers, we have spoken about dividing a whole into parts and recognized the individual reality of both wholes and parts. We also said that the whole is outside the domain of parts; although it can be a member of another domain of the parts of a larger whole. Rational numbers too invoke a relation between whole and part but if we interpret these as involving complex numbers, we cannot now treat them as integer fractions. This eliminates quantities, cardinals, and whole-part relationships—all ostensibly normal interpretations—from potential use here.

We have also interpreted negative numbers as concepts different from the positive numbers, which can both be concepts and individuals. This definition still stands, and I propose that we extend it. A rational number can be viewed as the attempt to interpret one idea in terms of another. If the division can be carried out without a remainder, one concept can be fully reduced to another. If, instead, the division leaves a remainder, then it implies that one concept can be approximated by another concept, but never fully reduced.

Think about the interpretation of a snake as a rope. Many of the traits of the snake are like that of a rope, and hence there can be an approximation that assumes that the snake is a rope. But there are also many differences, which means that if we treated the snake as a rope, we would be making an incorrect judgment because we leave out some of the aspects of the snake in thinking of it as a rope.

By the Fundamental Theorem of Arithmetic[24], every integer can be represented as a product of prime numbers. Since fractions are the division of such integers, they ultimately involve a product of prime numbers by another product of prime numbers. If the fraction reduces to an integer, then the numerator has all the primes in the denominator, and the division produces an ordinal—i.e. the sequence in which we can repeat an idea to produce another idea. But if the fraction doesn't so reduce, then fraction represents the failed attempt to interpret one idea in terms of another. It is an approximation that tells us the extent to which the attempt to interpret has failed. This failure is not necessarily a bad thing. For instance, if you were to think of a block of wood as a chair, you can indeed use it as chair. The approximation is not bad; it works in many ways as a chair. But is the idea of a block of wood identical to the idea of a chair? No. They are distinct ideas but sometimes you can approximate an idea with another and the approximation works for many purposes. And yet, since we cannot equate the two ideas, we have a remainder that represents why the two ideas are not reducible to each other.

Ultimately, this irreducibility is due to prime numbers; every fraction that is not a whole number is like that because the numerator has some primes absent from the denominator or vice versa. Therefore, ultimately, the attempt to interpret fails because of the irreducible ideas—the primes are the irreducible ideas that can't be interpreted in terms of other irreducible ideas. Since the primes are infinite, we are now left with a new problem, namely, that there are infinite irreducible ideas—something that I will try to tackle separately. This approach is Platonic in the sense that we are recognizing the reality of ideas, supposing that there is an infinity of them. We still need to ask: How did the Platonic ideas come into existence—i.e. how were they constructed? The simplest assumption in Platonism is that this world is eternal; while the multiplicands of primes can be constructed, the primes themselves can't. So, you could create new ideas using the primes, but not primes themselves. This problem is resolvable only if we find a method to construct the primes—something that remains an unsolved problem in modern mathematics, perhaps the greatest unsolved problem today.

The problem is not completely intractable. For example, by Goldbach's Conjecture[25], every prime greater than 5 is the sum of three primes. For instance, $7 = 2 + 2 + 3$. In one sense, the primes are multiplicatively irreducible but not additively irreducible. The issue is that all sums of primes are not necessarily primes. For instance, $6 = 2 + 2 + 2$, which is not a prime. So, all primes are sums of primes, but all the sums of primes are not prime. How do we distinguish the sums of primes that are primes from those that are not?

I will not attempt to answer this question here, but I can say that the ability to construct primes by adding primes resolves the problem with Platonism—that all ideas are eternally irreducible. We don't need infinite irreducible ideas; we just need three irreducible ideas—represented by numbers 2, 3, and 5—and we can construct all the other primes. If we construct all the primes, we can construct every other integer. From those integers, we can create the fractions. The only additional thing required apart from the primes is the existence of *operations* like addition, multiplication, subtraction, and division. As far as numbers are concerned, there are only three basic numbers—2, 3, and 5—and with Goldbach's Conjecture every other prime is the sum of these three primes. With the Fundamental Theorem of Arithmetic every number is a product of some prime.

So, if we treat numbers as ideas, we aren't necessarily left with an infinite Platonic world of irreducible ideas, provided we accept the existence of number operations also as a fundamental fact. It allows us to reduce the Platonic world to a handful of fundamental ideas and operations—which is important to construct successive ideas. The entire problem of number theory reduces to the understanding of these fundamental ideas and operations upon those ideas.

Irrational Numbers

In contrast to integers, irrational numbers have fractional parts. In fact, all irrational numbers have infinite digits in the fractional part. An irrational number can be expressed as an infinitely continued fraction as below. To specify an irrational number, we need to specify the succession of integer coefficients—a_1, a_2, a_3 ... a_n. Since there is an infinity of such coefficients in each irrational number, the program that specifies these coefficients would also have to be infinite[26].

$$a_0 + \cfrac{1}{a_1 + \cfrac{1}{a_2 + \cfrac{1}{a_3 + \cfrac{1}{...}}}}$$

Irrational numbers present several difficulties when they are interpreted realistically. First, we cannot think of these numbers as representing the cardinality of a set, because how can a set have a fractional number of members? Second, irrational numbers cannot be realistically visualized as locations in space because each such location would require an infinite amount of information. If nature limits the amount of information that can be encoded in a finite space-time, then irrational numbers cannot exist as points in that space-time. Third, when points are labeled with numbers, it seems that all points on the real line are not identical because some points need infinite information while other points need a finite amount of information. For instance, the point labeled 1/2 needs a finite amount of information while $\sqrt{2}$ requires infinite information. How can all points be identical if different amounts of information are needed to construct them? Fourth, the fact that an irrational number requires an infinite amount of information contradicts the fact that many irrational numbers can be constructed geometrically using a ruler, compass and isosceles triangle in a few steps. For instance, common irrational numbers such as π and $\sqrt{2}$ are easily constructed through a circle and a right-angle triangle. So, irrational numbers have a finite complexity when it comes to geometrical construction, although require an infinite complexity if represented arithmetically.

These interpretive difficulties require a deeper understanding of the difference between natural numbers and irrational numbers. This understanding is indicated by Georg Cantor's work on infinities. Cantor was studying the properties of various classes of numbers and he demonstrated that natural numbers are fewer than irrational numbers. This isn't surprising given that there are infinite irrational numbers between any two natural numbers. And yet it is hard to prove this fact because natural numbers too are infinite. To claim that there are more irrational numbers than natural numbers amounts to saying that one infinity is greater than another. This idea seemed outrageous when Cantor first proposed it, but it is considered one of the finest mathematical results today. The idea that infinities are bigger or smaller has remarkable implications for mathematics.

Cantor showed that natural and irrational numbers belong to two distinct sets both of which are infinite. However, the cardinality of the set of irrational numbers is far greater than the cardinality of the set of natural numbers. Cantor's proof revolves around showing that there is no method to map natural numbers to irrational numbers one-to-one. Cantor had earlier proved that natural numbers can be mapped one-to-one to fractions which is a surprising result because there are infinitely many fractional numbers between any two natural numbers and, naively, there would be far more fractions than integers. Cantor however showed that it is possible to map fractions and integers one-to-one but there is no method to do the same for irrational numbers. It followed that while integers and irrationals are both infinite, the infinity of irrationals is greater than that of the integers. If the cardinality of irrationals is α_1 then Cantor showed $\alpha_1 = 2^{\alpha_0}$ where α_0, is the cardinality of integers.

Cantor then suggested that infinite sets have different sizes and successively bigger infinities are exponentially larger than the previous infinity. Natural numbers are the lowest infinity; we could call this the countable infinity because it includes natural numbers. Irrational numbers constitute the next exponentially higher infinity. This means that there is no infinity between α_0 and α_1.

Cantor's ideas on numbers can be connected to the theory of sets, since a set of N objects has 2^N subsets. For example, a set of 1 object $\{1\}$

has 2 subsets—the empty set {} and the set with one element {1}. A set of 2 objects has 4 subsets—{}, {1}, {2}, {1, 2}. A set with 3 elements has 8 subsets and so on. As N grows linearly, 2^N grows exponentially and we cannot label the sets using natural numbers. The point is contentious because if N represents natural numbers then N is infinite, and we are talking about exhausting infinity. But Cantor's point was that if a quantity grows exponentially then it cannot be mapped to natural numbers that grow linearly. He defined the cardinality of a set as the total number of elements in it. A cardinal number represents the size of a set. It is now possible to speak of the cardinality of infinite sets. If we speak of the cardinality of the set of natural numbers, then it is a countable infinity. Cantor defined this cardinality to be α_0 (called aleph 0). If we create subsets out of the set of natural numbers and if there are α_0 natural numbers, then the cardinality of the set of all the subsets will be 2^{α_0}.

But what does the set of natural number subsets represent? Assume that we have a universe of objects and we have distinguished and labeled them with natural numbers. The universe can be divided into object collections. Since objects have been labeled uniquely by numbers, a collection of objects is also a collection of numbers. Subsets of the object collections set can denote number collections, and this allows us to transition from a theory about numbers into a theory about sets. From set theory we know that a universe of N objects, there are 2^N ways in which it can be divided. From Cantor's proof we know that the cardinality of the irrational number set is also 2^N. This means that the number of irrational numbers is the same as the number of subsets of the natural number set. It is now possible to give an interpretation to irrational numbers—namely that they represent collections of countable objects. Even though the objects can be counted, the collections cannot be counted; the methods of counting objects cannot be applied to sets as they are uncountable. The collection of all subsets of a set is also called its *power set* and for a set with N elements, its power set has 2^N elements.

Of course, the power set of N contains N which is countable. But it also contains elements that cannot be computed. These elements correspond to irrationals such as π and $\sqrt{2}$. This addresses all the above problems encountered while interpreting irrationals. First, we don't have to view the irrationals as points on a line; we can view them as collections of natural numbers. Second, if irrational numbers are not points then the information difference between a finite integer and an infinite irrational is not infinite. Third, this is confirmed by the fact that irrationals can be constructed using a finite number of geometrical steps. Fourth, this would clearly mean that the irrational number does not represent the cardinality of a set. Rather, we must give these numbers a new interpretation. Recall that numbers can be used as names to individuate objects. By extension, we could treat irrational numbers as names of collections.

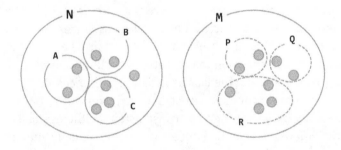

Figure-7 Numbering Object Collections

The infinite digits in an irrational number only arise because we are trying to count all the *possible* ways of dividing the world into objects, although only *one* such way is real. Figure-7 shows two possible ways in which objects can be counted, each time through a different set of hierarchies. In this scheme, the names of objects and collections are countable if we only look at one scheme of counting. The names become uncountable if all possible schemes of counting are used. Irrational numbers thus represent not the *real* state of the universe, but all potential states. If we call an object by its irrational number, we would be recognizing not just what it *is*, but also everything that it is *not*. If we look at the universe, it is countable. If we look at the possibilities, then it becomes uncountable.

A collection represents a relation between things. Even if all the objects are real, the relations between them remain possibilities. The reality of irrational numbers is that they exist as a possibility. With this view, can revise our view of numbers, yet again. We can say that numbers are ordinals, which are concepts, which are possibilities. Therefore, all concepts are existing, but as potentials. The natural numbers are potentials because we can potentially count till infinity, but infinite number of things may not exist. The complex numbers are potentials because there may be potentially infinite levels in a cascading hierarchy, but infinite levels may not exist. The rational numbers are potentials because we can interpret any concept in terms of any other concept, but the interpretation may not exist. And the irrational numbers are potentials because there may be relations between any two objects, but those relation may not exist.

Potentiality simply means that there is a path or process by which something can be constructed, but that process must be applied or used to produce the output of that construction. The existence of a number is contingent upon that construction. But otherwise, we can continue to speak about them as possibilities. Our focus should not be so much on whether numbers exist, but on the operations, processes, and methods by which they are constructed. Mathematicians focus upon the different classes of numbers but are not equally interested in all the methods of producing them because they think of these numbers as *objects* or things that exist, not something that exists only as a potential, which becomes real on the application of a process. The reason is that considerations about processes or operations compel us to think about the real world because the application of an operation involves force and energy. Bertrand Russell famously said that mathematics is about what is true in all possible worlds, not in the present world. But is this stance true? Should we be interested only in the possibilities, or also the methods by which they are realized? Isn't it likely that there are many more types of numbers than presently known, just because we may not have understood all the methods of creating them?

Are Numbers Quantities?

The idea that numbers are quantities forms the basis of nearly every scientific endeavor today. This idea works perfectly for objects (if objects are *a priori* distinct entities) because numbers can now be equated with the measurement of objects. Indeed, all physical sciences measure object properties and describe them using numbers. Given the myriad successes of these descriptions, it seems like a foregone conclusion that numbers are quantities. However, given the previous discussion, this notion about numbers fails quickly the moment we speak about negative numbers, because these cannot be measured: How can you measure the absence of something? Furthermore, how could you say that something is *more* absent when we speak about -10 and *less* absent if we speak of -5? This doesn't entail that numbers could not be used for measurement. It just means that numbers *aren't* quantities of measurement. Numbers represent a sequence, each member of that sequence and the sequence collectively represents a concept, and to create that sequence and its members we must employ operations.

It is more appropriate now to speak of numbers in two distinct ways—(a) as an idea, and (b) as a symbol. The symbol *represents* the idea, but it is a physical instantiation of that idea. The idea is potential, and the symbol is real. But to convert that potential into a reality you need some type of force, energy, and activity. Therefore, we can divide the study of numbers into ideas, actions, and symbols. The distinction between a symbol and an idea is very subtle: a symbol is an idea, but an idea is not necessarily a symbol. The difference is that the symbol has existence or reality, but the idea is simply a potential or possibility that hasn't yet been realized. We cannot produce a complex possibility from a simple possibility, because that complex construction involves an operation which will produce a symbol. Therefore, it is impossible to speak about numbers simply as possibilities. The oft adopted notion that mathematics can deal purely with ideas is false. To create complexity, we need operations and those operations will produce symbols. From those symbols we can produce further symbols. Even the existence of numbers in our minds is existence as a symbol, not as a possibility. Therefore, even if mathematics is viewed as a purely mental activity, it is mental effort that produces the symbols by converting a possibility. In short, we cannot deal just with ideas; we must deal with ideas, activity, and symbols. We need three rather than just one category.

Every activity creates, destroys, or preserves a distinction. Only when things become distinct do they become countable, and numbers can be applied to that distinct entity. Therefore, actions are the creators, destroyers, or preservers of distinctions. When we create a new distinction, we create a reality that was previously a possibility. When we destroy a distinction, we convert a reality back into a possibility. We can convert ideas into symbols—i.e. the conversion of a possibility into a reality. We can also convert symbols back into ideas—i.e. the conversion of the reality into the possibility.

There are hence three states of existence—possibility, action, and reality—and mathematics must deal with each of them. The subject cannot just be about ideas, and steer clear of the force and energy that produces a reality from a possibility. The three are necessary even to do mathematics. But once we realize that these three are necessary to do mathematics, we can also see that the real world is not different from mathematics. In short, there isn't a real world that we simply *describe* using mathematics. Mathematics *is* that real world, and the real world is also mathematics. The division between reality and mathematics is an artificial creation of the modern way of thinking about numbers because we treat numbers as a different kind of reality than the world in physical sciences. Then we are left with the paradox of how mathematics is so successful, or why matter is governed by logic and mathematics. If we dissolve the distinction between the two, then the real world and mathematics must both be studied using the three categories noted above.

Type Number Theory

In this chapter I have taken you through a journey in which we modified the definition of a number several times:

- Numbers are quantities – the commonsense view.

- Numbers are cardinals and ordinals but not quantities when we speak of them in terms of positive integers.

- Numbers are ordinals but not cardinals, as the cardinal view creates problems with negative numbers; instead numbers are ordinals associated with a direction or dimension. This makes numbers inseparable from the idea of space.

- Dimensions can be embedded inside objects; this creates a hierarchy partially denoted in complex numbers. This produces the notion of a hierarchical space.

- Numbers are ideas; prime numbers are irreducible ideas, and fractions are comparisons of one idea against another; these fractions represent our mental state of cognition.

- Numbers are possibilities, which arise when we posit relations between integers that are not always real.

- To convert the possibility into a reality, we need a force or energy, thereby converting an idea into a symbol.

As we discussed earlier, the physical nature of the symbol entails that we can only apply natural numbers to it. However, the *meaning* of the symbol can use all the other kinds of numbers—negative integers, complex numbers, rational and irrational numbers. That is, physically, a symbol is an individual object, and you can only count these objects from 1 to ∞. But what the symbol represents, can be understood in many ways. The symbol denoting our understanding of the world can, for instance, be represented by a fraction. The symbol denoting the whole-part relation can be represented by a complex number. The symbol that logically contrasts all that exists with all that is possible but doesn't exist can be denoted by an irrational number. In so far as a symbol represents something that is real—without invoking the possibilities that don't exist—it can be counted like a natural number. Hence, negative numbers, rational numbers, and even complex numbers (if they don't involve irrational numbers) can be counted like natural numbers. Only irrational numbers cannot be counted like natural numbers because they deal in the possibilities that don't exist but are conspicuous by their absence because we invoke their existence in the meaning.

Since all symbols are produced from ideas, the world is symbolic in the sense that it encodes ideas. Its physical existence can be known by a natural number, but its meaning can be known in more than one way. In other words, meanings are more exhaustive than the physical existence; in fact, the domain of meaning includes natural numbers. So, the meaning of a symbol can be a natural number (although it might be a different natural number than the physical existent). Beyond this, the meanings can be negative, complex, rational, or irrational numbers. Given this fact, we can ask: is the *behavior* of this symbol—i.e. its causal properties—defined by its meaning or by its physical individuality? I would argue that in dealing with causality we must always refer to the symbol's meaning rather than its physicality. In case the meaning and the physicality are the same natural number, meaning equals physicality. But that is an exception rather than the rule. This exception can lead to the erroneous conclusion that causality is given by the physical properties. Meaning in mathematics enters explicitly when we consider symbols whose meanings and physicality are not identical natural numbers.

Our ability to count symbols and our ability to count their meanings are therefore two separate things. They overlap in the case of natural numbers, but not otherwise. Owing to this fact, we can measure a kilogram of rice against a kilogram of gold, or a million pieces of paper against a million \$100 bills, and our ability to map them 1-1 indicates the quantity but not the meaning. The domain of meaning is bigger than that of natural numbers and includes natural numbers. So, if we study numbers as meanings, we will also study natural numbers but if we only study objects then we can understand natural numbers correctly, but the rest of the numbers will remain a mystery. This leads to my contention that numbers are ideas, which preexist things. When we convert these ideas to things, and study those things, while ignoring what they *represent*, we only capture the natural number portion of a semantic reality. But if we study these things as symbols then we can also understand the semantic reality beyond the physical one, which is more extensive as well.

This is what I mean by a *Type Number Theory (TNT)* where numbers are meanings. Firstly, it is a philosophical idea that answers the question "What is number?" consistently across the various types of numbers. As we have seen, many notions about numbers such as quantities, cardinals, or even ordinals fail when we try to understand different kinds of numbers. Mathematicians may carry on with computation and theorem proving without answering the above question, but at some point, everyone asks themselves:

What are we really studying in mathematics?

My claim is that we are studying *ideas*; in fact, we are studying different classes of ideas. Since numbers can be used in counting things, and studying their causal properties, every idea must be a number and every number must be an idea. Mathematics is therefore coextensive with all of philosophy, natural sciences, and even the study of the mind. We just haven't seen that connection yet, because we haven't found a consistent way to look at all kinds of numbers. We also have an anathema to the reality of ideas, at least in the physical sciences. So, philosophers have a tough time claiming the reality of ideas as things that may exist outside our mind. Mathematicians who accept this reality, posit that ideas must exist in another world of ideas beyond this world. But that doesn't help us in two ways. First, if they are in the other world, then how do they appear in this world—e.g. in our minds? Second, why are these ideas so useful in modeling the world when we suppose they are in a different world? Platonism is a flawed approach due to these problems. The right approach is to view the world symbolically rather than physically; it exists, but its existence is not the most important thing; rather, the meaning represented by that existence is more important.

Second, this meaning is both objective and relative. By objective I mean that it exists in the object as a *possibility*. By relative I mean that one of these possibilities is contextually selected to become a reality, and this selection involves a choice and a purpose. We cannot therefore attribute the meaning entirely to the symbol, or entirely to the contextual use. Philosophers in the West have been torn between either of these extremes. Some linguists believe that meaning is in the words or the sentences, so it is objective, and it can be derived from the dictionary meanings and grammar. They sort of forget that dictionaries and grammar are as much our creation, and they are not static either. New words are constantly added, old words change their meanings, and the use of these new and old words constantly evolves. Other linguists believe that meaning is contextual, which means there is no objective meaning in the words themselves, so dictionaries and grammar are useless. What truly matters is the conventions by which we use the words in different situations.

You can see how these two extremes are pointless. We must either throw out grammar and dictionaries and rely only on the context or throw out the context in favor of dictionaries and grammar but neither of these alternatives works. The problem is not unique to linguistics but exists in mathematics as well. Should we treat the numerals as universal significators of universal ideas, or should we just regard them as cultural methods of representing these ideas? When you extend these questions into the physical sciences, you find more problems because we cannot study the world of objects unless objects have some causality, and yet, we cannot attribute causality only to the objects (owing to atomic theory) because causality in the object acts occasionally and selectively in relation to some objects. To what should we attribute this selection of location and time?

These positions are easily reconciled by viewing the physical symbol as a collection of possibilities—which are objective in the sense that they rule out other possibilities (which are outside the collection)—and yet, the context selects a possibility. So, what you observe is the culmination of three things—(a) a set of possibilities, (b) a relation that selects the possibility, and (c) a choice that selects the relation. The possibilities are objective, the relation that selects a possibility is intersubjective, and the choice is subjective. In other words, the meaning of a symbol exists objectively in the symbol, but it is one of the many possible meanings you can attribute to it. Only context—determined by a choice—fixes that meaning.

It follows that numbers are not just *ideas* but also *possibilities*. An object is a *collection* of possibilities. And within a context one of these possibilities is instantiated to produce a meaning. Therefore, we can distinguish between three things—(a) possibilities, (b) the set of possibilities, and (c) the selection of these possibilities. The possibilities are universal meanings, a set of possibilities is produced by an action, and the selection of possibility requires a context and a choice. The choice in turn is not unlimited; it is limited by what is *available* for selection, by the goal to be achieved, and by the rational understanding of the most optimum way to achieve it. This greatly broadens our ontology to many distinct categories, but it also makes the system in which we are theorizing consistent and complete.

Thirdly, since we are dealing with possibilities, we must also embrace an *anti-reductionist* approach. In reductionism, we suppose that the small things combine to create bigger things. That is certainly true of the *collection* of possibilities where individual possibilities are combined to create a set. But that is not true of the members of the set themselves. Instead, a larger possibility *divides* into smaller possibilities through a process of restriction. The set of possibilities is combining to create a larger possibility, but the individual members of the set are produced by dividing a larger possibility. In modern mathematics, we only consider the former mechanism, and we are therefore led to the question of how the 'atoms' of the set are produced. Mathematics doesn't delve into this question; it supposes that some physical theory will deal with this problem of defining how the smallest objects of the world are produced from energy.

But we now know that these smallest subatomic particles are possibilities (through atomic theory). When they form an atom (e.g. Oxygen or Carbon) or a molecule (e.g. Sugar) a set or collection of possibilities is formed. This set can create different behaviors in different relations (which we call chemical reactions), and the individuality of the subatomic particles in an atom is never in question (through measurements). So, we are required to distinguish between two kinds of possibilities—(a) the member possibility, and (b) the act of selecting a member from a set or collection.

There is reduction in this scheme too; it involves the reduction of the collection to an individual. But we must remember that the individual is also a possibility because it is not always causally active or observed. Therefore, the member is a possibility, and the collection is another possibility. The collection can be reduced to member parts, but what do we reduce the members to? Reduction claims that these members must be reducible to smaller parts, which would in turn be reduced to even smaller parts, indefinitely. If reductionism were true all the way, then there would be no limit to how small you can divide a set into its members. But if reductionism has a limit, then we must uncover the reason why it exists.

Take for example a very large number. We don't say that this large number is unreal because it can be reduced to smaller numbers. We claim its reality, quite like the reality of the smaller number. If both these numbers are possibilities, then we are acknowledging the reality of a large number as much as that of the small number. The anti-reductionism begins in the claim that the big is as real as the small. It has an identity or individuality like the small. Therefore, there can be big ideas and small ideas, big possibilities and small ones. This acknowledgement defeats the basic premise that only the large is real, however, it doesn't say if the big or small could be mutually reduced. For instance, the class of mammals comprises dogs, cats, cows, horses, etc. Should we say that the notion of a mammal is meaningless because it is *only* a collection of parts? Or should we say that the idea of a mammal is as real as that of cats, dogs, cows, and horses—because it can be defined as those animals which breastfeed their children? Similarly, is the set of rational numbers simply a collection of member parts, or is the term 'rational number' meaningful in some way that we don't have to know all the member parts to know what the set itself represents?

Clearly, in all practical terms, we are anti-reductionists because we give reality to sets, beyond the reality accorded to the member parts. However, the nature of this reality is not *physical*—i.e. an object in the conventional sense of members. We accept that the set represents an idea, and because the idea is meaningful, it doesn't have to be enumerated in terms of its parts. In fact, certain sets—e.g. 'irrational numbers' cannot even be enumerated. So, what does the term 'irrational number' mean if it is irreducible to parts?

But I would even go beyond this type of anti-reduction in which both parts and wholes are real. I would rather claim that the part is *derived* from the whole because in practice you can never enumerate all the parts for any reasonable type of concept. For instance, you could take the idea 'dog' and try to enumerate all the dogs in the world. But does this set also contain dogs which existed in the past and will exist in the future? How do you enumerate those members which you cannot even observe? It is more appropriate to say that the idea 'dog' always exists, but it *instantiates* into individual dogs at different places and times. To define 'dog' I cannot use enumeration because it can never work. I must either say that 'dog' is meaningless or find another way. That way is to accept the reality of dog, even apart from the members, but give it *priority* over individual dogs. In short, the idea precedes the individual instantiations. This is a more radical type of anti-reduction than simply giving reality to both objects and the ideas they represent; we are now claiming that the individuals are produced from the idea, because the only alternative without this premise is that ideas simply do not exist. If ideas don't exist, then sets are unreal, and mathematics is impossible. So, by TNT I don't just mean that we are studying ideas in mathematics, but that ideas have priority over things that represent these ideas.

Thirdly, we can now further extend this anti-reduction even further by stating that the bigger ideas are the origins of the smaller ideas. In short, we are beginning with the biggest possibility—everything—and reducing it gradually to produce something specific. It remains a possibility but both big and small possibilities are distinct *individuals* rather than *composites* of the smaller individuals. So, just like I cannot say that 'dog' is the enumeration of all dogs, similarly, I cannot say that 'mammal' is the enumeration of all dogs, cats, horses, cows, etc. Just as 'dog' instantiates into individual dogs, similarly, 'mammal' instantiates into individual types of mammals. By this anti-reduction, the biggest idea precedes the smaller ideas. The *simple* is now the *biggest* and the *complex* is the *small*. Mammal is therefore not a complex idea produced by the combination of cat, dog, horse, cow, etc. because it is possible there are other species which may have existed in the past or may exist in the future, which breastfeed their children. Since I cannot enumerate them, therefore, I must accept 'mammal' as a primitive idea which gets instantiated into other subtypes in different places and times. The world doesn't begin in the smallest things and becomes complex through an aggregation of these things. It rather begins in the biggest ideas which get subdivided into smaller ideas to produce complexity.

We can now apply this anti-reductionism to numbers by saying that the larger number is produced from the smaller number through a restriction. Like the reductionist produces a bigger thing by adding smaller things, it is possible to invert the process by taking the biggest thing and restricting it through successive steps.

My claim is that 0 is the biggest possibility. It can be equated to *nothingness* because it cannot be observed, except through a choice and relation that selects everything at once. All numbers proceeding from 0 are smaller possibilities, and the largest number—∞—is the smallest possibility. You can imagine this type of construction by looking at M. C. Escher's drawing shown in Figure-8. The 0 is the origin or the center of this picture, and you can count outward from this 0. You initially get large parts, but as you continue the process, the parts become smaller—in fact infinitely smaller. There are infinite parts of this picture—an angel or a demon—but infinity is now *bounded*. You can never reach infinity by counting, but that doesn't mean it is boundlessly far from the origin. The reason is that your successor ordinals are smaller than the predecessors.

Figure-8 Escher's Drawing of Angels and Demons

Therefore, as we count from 0 to ∞, we are not counting from a small number to the biggest number. We are rather counting in reverse from the biggest idea to the smallest idea. It just seems that ∞ is big because it is arrived at by a long succession of steps; cardinally and ordinally it is big, but quantitatively it is the smallest. Quantitatively, infinity is the infinitesimal. It takes us infinite steps to construct the infinitesimal, each of those steps is an ordinal, the sequence of all those steps is the cardinal, and the result is the infinitesimal. We begin from the biggest idea—and that idea is 0. We then restrict it to produce smaller and smaller ideas, and if nature has a limit to that restriction, it is because beyond a point *we* cannot distinguish. It is a limitation of our conceptual and perceptual apparatus, not a limitation in principle. The limit on reducibility makes the universe finite, although it can have infinite parts. Basically, the biggest idea or possibility is itself finite, or has a *form*. But it is also infinite because it can have infinitesimal parts. The basic contradiction between finite and infinite hence doesn't exist.

Fourthly, we can ask ourselves: What is a symbol? Recall that we form collections of possibilities, which constitutes a symbol. These possibilities are themselves produced by restricting the whole, so they are subsets of the whole, but they can be further subdivided by a choice. The whole is therefore the original idea; its subdivision by an act of selecting from the whole is the symbol, and its further subdivision through relations and choices is the contextual meaning of the symbol. The construction from the original idea to the symbol with a meaning involves successive restrictions. A symbol is the original idea with an individuality. And the symbol's meaning is that symbol in a relation. We just apply these restrictions in different ways—the former is the act of a selection within a whole, and the latter is the act of selection through relations to other parts.

To think in this way, we need three categories—the universal, the individual, and the relational. The ideas are universals. But their instantiation into a thing involves individuality. Finally, their relation to other things creates a meaning of that individual. As we spoke earlier, we can also designate these three categories as the objective (universal), subjective (individual), and intersubjective (relational). We can distinguish between an *idea*, which exists universally, and an individual instance of that idea, which is the symbol. Then we can distinguish between the symbol, which combines the universal and the individual from the *meaning*, which is given contextually.

The process of constructing the parts from the whole forms an inverted tree of ideas, which I will call a *semantic tree*. It is a tree of universals in which the root is the biggest or the most complete idea. Parallel to this tree is another tree of individuals which comprises of parts of the whole, but this time the whole and the part are defined as things rather than ideas. The elements of the tree of ideas combines with the elements of the tree of individuals to create symbols, which are both individuals and universals. We can say that they are concepts and properties of a thing as understood independent of other things (although they must be viewed as parts of the whole both as individuals and ideas). These symbols are then given contextual meaning through relations. For instance, there is the universal idea of a chair. It can be combined with individuality to create an individual chair. But in a context, this chair can be used as a table. Similarly, there is the universal idea of a dog. Then there can be an individual dog that instantiates this idea. But, contextually, a dog can act as a horse if it pulls a carriage. Therefore, there is a very important role for contexts in defining meaning. But the substrate on which we define these relational meanings is also idea-like. It just seems to be a thing because an idea and individuality have been combined in producing that thing. That idea is the *possessed* property of that thing, while the meaning is the *relational* property.

5

Mathematical Foundations

In the sky, there is no distinction of east and west; people create distinctions out of their own minds and then believe them to be true.

—Gautam Buddha

Mind, Matter, Mathematics

We have a poor understanding of the relation between mathematics and reality on the one hand, and mathematics and the mind on the other. Mathematicians view mathematics as the study of what is possible, not what exists or what we understand. That has prompted a separation between the study of abstract structures in mathematics and the study of matter and mind in other sciences. The problems of semantics require a relook at this approach to mathematics because problems of semantics appear through the nature of numbers as concepts, which are used both to study nature and the mind. The problems further appear in dealing with different interpretations of numbers as names, concepts, programs, etc. because mathematics is unable to deal with meanings. The inability to deal with sets as ideas which stand on an equal footing as the member objects, or the inability to treat ideas as logically preexisting objects, or as meanings that are expressed through relations, or that symbols combine universal ideas and individual things, all contribute to the problem. When mathematics is rooted in objects rather than ideas, everything else that follows from this root also becomes non-semantic.

There is also a more generic problem concerning the nature of the mind. The problem is that science has described the world in terms of mathematics, and we would like to extend that ideology to minds, formulating mathematical theories of the mind. For instance, we can envision machines that think like human beings. These machines would be described in a mathematical theory. How can we imagine the possibility of thinking machines when the mathematics in which these machines are described is itself incapable of meanings? To solve this problem, mathematical foundations must be construed in a manner that can produce meanings, not just tokens. Such a mathematical theory will also be a theory of the mind.

Finally, there is the problem of how physical objects themselves are distinguished and counted. Physical theories assume *a priori* distinct objects, and mathematics inherits this assumption as well. How can a theory that assumes the existence of distinct objects explain the origin of objects? We would have to assume the universe has always existed as a set of distinct objects, and there cannot be a mathematical theory about the origin of objects. Note how this problem is directly a problem of physical sciences, even without an explicit reference to the nature of meanings in language or the mind. But in so far as physical sciences require a mathematical theory, the question about the origin of objects is also a mathematical problem. The idea that objects are *a priori* distinct fails to answer the origin of this distinctness. If, however, objects are not *a priori* distinct, then we must seek their distinctness in our ability to draw distinctions.

Foundational problems in mathematics therefore arise in three different ways. First, they arise because of the need to encode meanings in language. Second, they arise because of the need for a formal theory that can explain the workings of the mind. Third, they arise because physical sciences need to explain distinguishability.

In the preceding chapters I sketched an outline of the approach that can fix these problems. The approach says that mathematics should be a theory that explains numbers as ideas that are instantiated as symbols and given meaning through relations. A set now is the division of an idea into smaller ideas. Order amongst these ideas indicates a succession by which some ideas are logically prior to other ideas. Structure represents relations between ideas which select one out of many possible ideas within a set. And space-time is the hierarchical domain in which ideas and order are expressed. We also need to imbibe three categories of individual, universal, and relational to understand the nature of symbols and meaning.

This approach will resolve the three problems in mathematics described above. First, it will provide a theory of how meanings can be encoded in language. Second, mathematics can be a theory of the mind. Third, it will explain the *logical* origin of objects. With these benefits, such a foundation is worth the required effort. Of course, in one sense, this approach is clearly more difficult because we must now speak about order, sets, structure and change without speaking about objects being ordered, collected, structured or changing! If mathematics seemed abstract, this new approach would make it harder. But in another sense, this could bring mathematics much closer to intuitive approaches that are frequently used in everyday life. For instance, a set must now denote a concept and subsets will represent restricted concepts. Structures will denote relations that create meaning, and order will denote a hierarchy. The intuitive base for conceiving a mathematical foundation independent of objects therefore exists in ordinary language and everyday experience, and that could help in the formalization of these intuitions.

A Critique of Set Theory

Mathematics uses two notions about numbers—cardinal and ordinal. Cardinals represent the *size* of a set or collection while ordinals represent the *order* of objects in that collection. For instance, if we have a set of 5 objects, the cardinality of the set is 5, while objects in that set are labeled as 1, 2, 3, 4, and 5. From the standpoint of counting, cardinals depend on ordinals because the only way we can know that a set has 5 members is if we have counted them as 1, 2, 3, 4, and 5. It would seem, therefore, that ordinals are more fundamental than cardinals since cardinals are known after ordinals. However, the order among objects depends on a method of ordering, which is used to draw distinctions between things before they can be ordered and counted. This method of ordering is called the Axiom of Choice as we saw earlier. By relegating it to a 'choice', we fail to investigate the active role concepts play in distinguishing.

To avoid dealing with concepts explicitly, mathematicians define numbers based on cardinality or quantity. Bertrand Russell, for instance, defined a number as the set of all sets that have a certain specified cardinality. Thus, the number 5 is the set of all sets that have 5 members. The problem here is that to know that there are 5 objects we need to be able to distinguish them before we can even count them. Distinguishing in turn requires an understanding of the *modes* or *methods* by which objects can be distinct. This distinction can be based on physical properties or on meanings. But since mathematics doesn't deal in the physical world or in meanings, both approaches take us out of the domain of current mathematics. For instance, to know that there are 5 members in a set, we need to be able to distinguish them before we can know that there are 5 and not 4 or 6 members. That in turn requires the application of physical properties or concepts, both of which are outside of current mathematics.

The remedy to this problem is the tripartite distinction between individuals, universals, and relations. A symbol is a combination of an individual and a universal; it already possesses a meaning, even before we relate it to other such symbols. The symbol therefore embodies two ways of distinguishing—(a) its physical individuality and (b) its conceptual distinctness from other symbols. Those are the two things we said above are missing in current mathematics.

So, even without contextual meanings arising from relations between symbols, mathematics is incomplete because it doesn't view itself as the study of symbols. By obsessing about objects, whose distinctness is never questioned, we create the above two forms of problems. The problem, of course, gets much worse when we bring in relations and contextual meaning. We can cast this problem very formally in terms of logic, and its three principles, namely, identity, non-contradiction and mutual-exclusion. What I have referred to previously as individuals, corresponds to the principle of identity. What I have referred to previously as universals corresponds to non-contradiction (a symbol cannot be a thing and its opposite at the same time). And what I have referred to previously as relations corresponds to the principle of mutual-exclusion. The difference now is that we are looking at each of them in terms of *distinctions*.

The physical distinctness of a symbol, for instance, isn't to be taken for granted. It must rather be understood as an outcome of the process of distinguishing individuals. We might distinguish things based on their location in space or time, but this brings the nature of space and time as the modalities of distinguishing things. If sets are formed by collecting objects, but objects require space and time, then space and time are prior to objects, which are prior to sets. It would make set theory not a fundamental theory in mathematics if it relied on a theory of space and time; rather the theory of space and time would be more fundamental because it distinguishes objects. Once objects have been distinguished, they can be collected in arbitrary manners and set theory could deal with the nature of collections. But the distinctness of objects precedes their behavior in collections.

Indeed, with the necessity of individuality, universality, and relationality—each of which involves distinctions—there is a necessity to speak about *three* kinds of space and time. They make individual things, universal ideas, and contextual meanings distinct. Once this distinction of symbols has been established, then we can speak about their collections. However, a random, unordered collection of symbols is meaningless. We also need order *within* that set to produce a *proposition* which orders the symbols into a sentence. The cardinality of this set simply tells us how big or small the sentence is. But the order—together with the symbols and their three properties—make the collection of symbols *meaningful*.

Once we have formed a meaningful sentence, then we can speak about whether it is *true*. The possible orders within a collection are regulated only by a 'grammar' which produces different meaningful sentences. But the truth of the sentence depends on the comparison of meanings—between that of the sentence and some axioms. In other words, some grammatical constructions can be false.

Set theory is hopelessly inadequate in dealing with the broader set of problems. First, it is not a fundamental theory because it assumes the distinctness of objects, rather than delving into the origin of the distinctness in a theory of space and time. Second, it totally avoids the discussion of meaning; the members of a set are objects rather than symbols, so their physical distinction apart from the concept they represent is never brought into focus. This becomes the root of the paradoxes in mathematics where both the meaning of a proposition and its physical identity are referred to. Third, the notion of a collection of objects as representing a concept is always incomplete because we cannot enumerate these members. We just imagine that we can collect all members of a certain type without knowing what those members are. Fourth, even if you collect members, without the order among them—which requires a method of ordering—the collection itself remains a possibility. If we treat this as a collection of symbols, its meaning is undetermined. Fifthly, set theory doesn't delve into how it is possible to order these members in different ways, thereby producing different propositions, and relations to other symbols can be involved in this ordering. The relation to a symbol isn't therefore merely something *between* two symbols but is in fact involved in changing each member involved in the relation. Sixthly, once you form a semantic proposition after selecting an order among the symbols, there is no way to determine the truth because we are unable to determine the meaning.

There is also a fundamental question about whether the set or the object is logically prior. In the current view, a collection of cars represents the idea of a car. But to collect the members of this set we must have a definition of a 'car' to identify only those things that are cars. This produces a circular dependency between the member and the set—we must have the definition of a 'car' before we start collecting them into a set, but that definition is obtained only after the collection is complete—which in practice is never achieved.

In so many ways set theory is an inadequate theory to form the foundation for mathematics. Current set theory is the generalization of a *class*, in which a class comprises of a collection of members. We equate this class to concept incorrectly, because a concept has an *intension* or meaning while a class only has an *extension*[27] or members. Furthermore, the members of this set—especially those members which are not in turn classes, are meaningless entities, and they are unordered, because of which you cannot use a set to formulate propositions, determine their meaning, or their truth. There is also a dual treatment of classes: as a collection of members they denote a concept, but as members of a higher class they are objects, on the same footing as the meaningless members that are not classes. This leads to the confusion that a set is sometimes a concept (when it has members) and then an object (when it is a member). This dual treatment of the set then leads to set theoretic paradoxes.

A Critique of Number Theory

Frege defined numbers as the cardinality of sets. This notion of number is wrong on three counts. First, it is possible that the objects in the set can be further divided into still smaller parts, thereby increasing the cardinality without changing the *meaning* represented by the set. If a number represents the meaning of the set through its cardinality, then that representation is wrong because the cardinality can change without a change in the meaning. Second, two sets that have the same cardinality can have different meanings. The collection of ten shoes and ten cars has the same cardinality but not the same meaning. By equating the meaning of the set with its cardinality, we are counting objects after stripping them off all meaning. This works if the objects being counted are point particles, which don't have any structure. But if the object being counted is in turn a set, which can be divided into parts, then the set's meaning cannot be understood if we only count it as an object without considering its meaning. Third, Frege's scheme assumes that the objects in a set—e.g., two shoes or two cars—are already distinct, without defining how they are distinct. By assuming the existence of objects that are distinct regardless of how we distinguish them, we presume an ontological objective reality. The knowledge that there are N distinct objects depends on the ability to distinguish and order those objects, which assumes counting. How can a theory of numbers assume counting?

Frege defined the number 2 as the collection of all sets that have two elements. In case you object that the definition of '2' depends upon knowing that a set has 2 elements, Frege said that we do not necessarily need to know which sets have two elements if we can define a proto set with two elements and other sets are mapped one-to-one to the elements of the proto set. For example, the proto set for the number 0 can be the empty set {}. The proto set for the number 1 can be the set {0}; the proto set for the number 2 can be the set {0, 1} and so on. Frege however assumed that a set comprises distinct things, in a way that is outside mathematics. In the case of a proto set, the distinct things are labels (1, 2, 3 ...), and their distinctness is based on some properties that make these distinct. Even if we assume that the successive numbers can be computed using a successor function, the definition requires computation on a primitive object (e.g., 0) which must be a distinct entity to begin with. If there is one distinct object, then we can assume other distinct objects. But the question is—how do we know that there is one distinct object, without presuming an *a priori* world of distinct, individual objects? Such a scheme employs a cyclical definition; it assumes distinct objects and the ability to count them simply maps the distinguishability of one set of objects into the distinguishability of another set.

The scheme is unsatisfactory if we must find the ways in which things are distinct. The way we count things should be an outcome of how things are distinguished, because that changes the cardinality of a set. If counting depends on the ability to divide and classify things before they are counted, then the assumption that there are *a priori* distinct things which can be mapped to another *a priori* distinct set of things is flawed. Both these sets of things are produced by the ability to count and mathematics requires a deeper connection between physical distinctness and logical distinctions embodied in numbers. This connection can be provided if the distinction between things is produced from the ability to distinguish things.

The Four Worlds

A symbolic understanding of everyday reality leads us to the view that mind, matter and mathematics are not separate worlds. The mind grasps ideas, but those ideas are independent of the mind. These ideas appear in the mind like symbols, which means that they are combination of the universal and the individual. The mind has its own individuality—we might call this the 'personality' of the thinker. This 'personality' selects certain ideas from the universal ideas. Owing to this individuality, only certain ideas will appear in a certain kind of mind. This may make certain forms of creativity impossible for certain people, or for that matter, even understanding certain ideas because their personality is not suited to incarnate the ideas. The ability to create or understand certain ideas—i.e. advent them in the mind as symbols—therefore depends on the individual.

Matter too is symbolic, and it represents ideas; however, there is a difference between the symbols we can think and the symbols we can perceive. This difference is the extent to which a symbol is a certainty. The perceivable symbols are more certain; the mental symbols are more possible. This distinction is one of relative possibility or certainty. But how do we make this distinction between greater or lesser certainty? The answer must come from the nature of individuality, which must be greater or lesser certain. This relative extent of possibility or certainty is different from the possibility or certainty of ideas. For instance, the idea 'table' is more certain than the idea 'furniture', but less certain than the idea 'kitchen table'. However, the idea of 'table' in the mind is less certain than the 'kitchen table' you can perceive. So, ideas have greater or lesser certainty, but similarly, individuality must also have greater or lesser certainty. We have already noted that universality and individuality are distinct. The fact that both are less or more certain, allows us to view the physical reality as ideas, even outside the mind.

This is important because of the Cartesian mind-body duality in which the physical world was devoid of ideas that existed in the mind. This dualism was conceived as the distinction between two 'substances'—i.e. they are equally certain. We can instead define the mind as the uncertain individuality of abstract (uncertain) and contingent (certain) ideas, and the body as the certain individuality of certain (contingent) or uncertain (abstract) ideas. Thus, for instance, a dog has the individuality of an animal within it, but you cannot see the 'animal' in a dog. You can however mentally perceive it, so there is nothing wrong in calling a dog an animal. Conversely, we can think about 'dog' and 'animal'. In the two cases, the universals are the same, but the individuals are different. This means that mind and body are not two substances; they are both individuals—the body is greater certainty and the mind is greater possibility.

We might note here that the mind is capable of perceiving possibilities while the senses are capable of perceiving certainties only. For instance, you can stand on a vacant land and imagine how it could be populated by a building with people within it. This imagination is also a perception, just not sense perception. Similarly, a mathematician can imagine the use of numbers, or their properties, without realizing them in the physical world. This imagination is also perception, just not the perception of the five senses. So, to say that the mind is perceiving is to say that it is seeing a possibility. Hence, we can say that mathematics is the science of the possible, but it can be applied to the perceivable real world, because these are just two different kinds of individuals—one less certain than the other. It requires us to take the leap in thinking that *possibility* is also an individual. It exists as a thing that we cannot perceive by the senses. However, the mind can perceive the existence of possibility. By mental perception, we can see physically non-existent ideas.

We might note here that possibility is different from *probability*. Probability is about sense perception—i.e. the times when we can see something. Since there are times when you don't see, probability simply expresses the times when you see or don't see things. Possibility, however, never disappears, so it can always be seen. In that sense, we cannot say that probability *exists*, though we can measure the occasional existence and compute probability. We can however say that the possibility always exists and is hence real, even though we may not necessarily observe it by the five senses.

This fact is most clearly illustrated when we say something is *impossible*. Clearly, no amount of measurement can confirm the impossibility of things; it may just be that we haven't yet observed it. In that sense, impossibility can never be sensually perceived. But you can mentally perceive the impossibility and prove that through reason by showing that it contradicts what already exists. The assertion of impossibility illustrates that we can mentally perceive the existence and non-existence of possibility, although what is possible may not be measured or observed by the five senses. This compels us to recognize the individuality that is possibility. This is important because we are now accepting two kinds of existents—those that are certain and those that are possible. We can also say that there is an abstract form of existence that constitutes the mind; we can force its manifestation as a probability, thereby producing some evidence that it exists, but that empirical observation is not the mind. The mind is the thing that persists as a possibility.

Once we recognize how the mind perceives mathematics, we can then say that this perception is a mental existence, but the ideas of mathematics are separate from the mental existence. Mathematics itself is the study of universals. These universals can appear as possibility in different minds or can be written down on paper as individual symbols. These mental and physical manifestations of the ideas don't change the fact that these are all symbols of ideas. As universals, it doesn't matter if we are thinking about mathematics or writing it down using paper and ink. We can treat it purely as the study of universals, regardless of how it appears physically.

In a sense, mathematics is the study of the pure Platonic world of ideas, although the ideas are not independent of each other. They are also organized in a hierarchy—from less certain and more abstract to more certain and less abstract—forming a tree structure. The separation between the universal and the individual means that these mathematical ideas can be applied to individuals, because the individuals are symbols that combine universality and individuality. In one fell swoop we can solve the mind-body problem, make mathematics independent of the physical world, and yet explain why mathematics is useful in understanding the physical world.

In one sense, there are three worlds—the Platonic world of ideas, the mental world of these ideas symbolizing in the mind, and the physical world that embodies these ideas. In another sense, there are just two worlds—the world of pure universals and the world of individuals that create symbols from the universals. We could call these the Platonic and the 'real' world, with 'reality' encompassing both the mind and the body. The latter position is closer to Platonism, but without the contradictions of the mind-body problem.

Once we recognize the separation of universals and individuals, we can now speak about a third kind of world of relations. We can call these relations *grammar*, as they are used to form propositions. In creating a proposition, a symbol is given a *role*—which we call figures of speech, such as noun, verb, adjective, adverb, etc. The grammatical structure of a sentence is different from the *content* of the sentence; the same structure can be used to say different things. For instance, the structure of the sentence 'this rose is red' is identical to that of the sentence 'this building is tall'; in both cases, a noun is attached to an adjective; but they make different claims. The ability to use the same structure to convey different messages has led some linguists to draw a distinction between 'deep' and 'surface' structure. Their claim is that languages are similar in their deep structure, and this structure must therefore be 'universal'. It is the basis of linguistic structuralism in modern time. However, the structures are different from the ideas embodied or represented as propositions through these structures. A language is not simply the grammar; it is also the vocabulary, or the symbols and ideas used in expression. These two in fact constitute complementary aspects of language.

Thus, by relations I mean the assignment of roles to symbols. The material objects are symbols, but they can be arranged in different ways—producing different orders—that form a different meaning. We have earlier spoken about the emergence of meaning through relations; we said that a symbol is a possibility, but a relation selects one aspect of that possibility to produce a meaning. Thus, meaning arises in a context, even though ideas exist even without these contexts. The former is contextual, and the latter is universal. Due to the context, the same word or sentence can mean different things in different contexts, which we call its *interpretation*. The basis of these interpretations or contextual meanings is relation. So, we can use many different terms like structure, context, meaning, or interpretation, and they are basically indicative of the same type of reality that is not perceivable as things but as relationships.

These relations are formed between symbols, which combine the universal and the individual. So, the production of meaning involves three worlds—the individual, the universal, and the relational. To these three worlds, I will now add a fourth one—the world of *logic*. Logic is also relations, but it differs from grammar. With grammar we create propositions, but with logic we create relations between propositions that are not propositions. We call these *consistency, contradiction, complementarity*, and *completeness*. Some propositions may be mutually consistent or contradictory. Others may be complementary and together represent completeness.

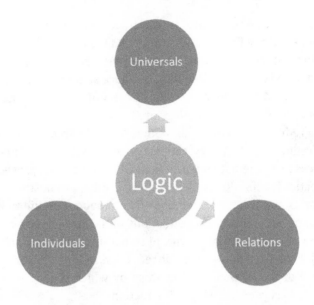

Figure-9 The Four Worlds

Whether we are thinking about mathematics, or writing down mathematics on paper, we combine these four worlds. First, we embody ideas into symbols, that combine universals and individuals. Second, we form propositions through relations. Third, we connect these propositions through logic to create theorems and proofs. If we recognize these worlds, then the individual numbers are concepts, that are denoted by numerals as individuals, they have a relation between them as successor and predecessor (although this relation must be understood structurally rather than merely linearly), and they can be related through logic into theorems and proofs. The division of mathematics into four worlds provides a clear conception of how mathematics is real, even outside the mind and the real world, and yet how it appears in the mind and is useful to the world.

Information and Reality

The problem in any foundation is that if something is fundamental, then it must also be irreducible. How do we know if something is irreducible? One possibility is that if we have reached fundamental ingredients, any further reduction will fail because the most basic entity isn't reducible. But how do we tell the irreducibility of the fundamentals apart from our inability to reduce something to even more fundamental? We could be wrong about fundamentals and might think that non-fundamental things are fundamental.

In the reductionist approach to knowledge, life reduces to biology, biology reduces to chemistry, chemistry reduces to physics, and physics reduces to mathematics. Mathematical foundations represent what mathematics itself reduces to. And this is a difficult question because reduction is not about *size* anymore. The reduction of life to biology to chemistry and physics are reductions in the size of matter. But the reduction of current mathematics to something more fundamental is a logical reduction, not a physical reduction in size. This breaks the continuity of the methodological steps by which reduction is done: reductions in size become logical reductions.

This problem can be fixed if physical reductions are also seen as logical reductions, which can be conceived in mathematics if the objects are treated semantically. A large physical object represents a complex meaning while a small object represents an elementary meaning. The limit to divisibility in matter ends with the limit of conceivability. That is, the point at which we cannot *think* of anything more fundamental is also the point at which we cannot divide matter into smaller parts. Counting is an outcome of dividing into parts. It is therefore an outcome—i.e. a phenomenon—of distinguishing, which is the cause. Reductions in the physical and logical worlds are reconciled if reality is *meaning bearing symbols*. The cardinality of a set approximately reflects the amount of information in it. A small set carries less information than a large set, and as we pack more information, the cardinality of the set grows. We must also remember that ideas are different from information. Ideas begin in the abstract and get detailed by refinement. The abstraction represents the 'bigger' ideas, while the refinement denotes 'larger' information. In one sense, there is no logical limit to divisibility of the larger ideas; the limit is physical and mental—i.e. in the types of individuals we can create physically or think mentally. In another sense, every individual is part of the biggest idea—'everything'—and contains it. The biggest idea is therefore present in every simple or complex information set, because information is about the number of individuals, while knowledge is about the universals. As knowledge decreases—i.e. as we speak about the ideas whose domain of applicability is limited—information increases.

In classical reductionism, we don't know when we have reached the point of no further breakdowns. But the semantic view helps us see that the smallest possible length in space and duration in time is also the representation of the smallest division. Therefore, the limit to logical reduction is the limit to physical reduction, and the limit is characterized by the smallest possible *extension* and *duration* in nature—that can be carried out physical or mentally. However, to see that this smallest possible extension or duration also denotes the limit of our senses and mind, we need to view the world as ideas.

Simplicity here represents the amount of information and not the ideas. For example, the concept of a table is simple. As this concept is refined and more details are added, we can come with a design for a table, such as the number of legs, the shape of the top, whether it has drawers, etc. This design is more detailed and hence more complex than the concept 'table' but it is not as complex as the real table. As a next step, therefore, there can be a prototype of a table that miniaturizes the design into a demonstrable artifact. This artifact has more information than the paper design, but less than that of the real table. Finally, one may elaborate the prototype into a real table in accordance with the design. Successive steps from the concept 'table' to a design, prototype and the real table are incremental stages of adding information. The conceptual table is simple while the real table is complex. Complexity can be built from simplicity.

In my book *Quantum Meaning* I have discussed a semantic interpretation of quantum theory, which treats atomic objects as symbols of meaning. This description of atomic reality overcomes the myriad problems of quantum interpretation such as uncertainty, probability and non-locality. In this interpretation, atomic objects are produced by distributing energy within an ensemble. Atomic theory permits many distributions, each depicted by an eigenfunction basis, and the theory does not predict which distribution is real. It is the measurement procedure that picks an eigenfunction basis. This view of atomic reality raises questions about the nature of an ensemble. Classically speaking, the atoms are real, and ensembles are imaginary boundaries that separate some particles from others. From a semantic view, an ensemble is a symbol—although a primitive one—like the word 'table'. As information is added to this primitive concept, we get symbols of much greater complexity. The ensemble is the 'root' of the real object which then springs from this root by the addition of greater information. Thus, we can think of every object as structurally an inverted tree—root above and leaves below.

If we never added information to the ensemble, it would remain an elementary idea without refinement. But the symbol representing this idea need not be physically bigger than the symbols that are subsequently added to it. Remember that the form of the symbol is different from the idea it represents. Therefore, the symbol can be small but the idea it represents can be abstract or big. Thus, as the branches and leaves spring from the root, a singular object—i.e. the root—becomes a collection of objects (the whole tree). And yet, all these parts were in one sense embedded within the root. This is the reason why large amounts of information can be *summarized* semantically. Popularizers of science present complex ideas in a simple manner to the lay public. Over time, as the interested reader pursues the subject, they add details to the original idea. The simple and summarized picture is not wrong; it is just not detailed.

This changes how we think of ideas, things and collections. Every idea can become a thing by adding an individual. This individual can then expand into a collection of individuals by adding details. In fact, this is the process by which an idea 'table' becomes a real table. The mind perceives that idea present in the real table, while the senses only perceive the details of that idea.

Logic as Distinctions

Classical logic is comprised of three principles called *identity*, *non-contradiction* and *excluded-middle*. Simply put, the principle of identity is that an object is always itself or that A is A. The principle of non-contradiction is that A and its contradiction (not-A) cannot be simultaneously true. The principle of excluded-middle means that either A or its contradiction (not-A) must always be true and both A and its contradiction could not be false at the same time.

The principles of identity, non-contradiction and excluded-middle cannot be reduced to other more fundamental ideas. To speak about these ideas, we must use all three ideas together. For instance, a discussion of non-contradiction can say that a thing cannot be simultaneously true or false, but it assumes that true and false are the only logically possible conditions, which is the principle of excluded-middle. A description of excluded-middle can say that true and false are the only possible conditions, but it assumes that true and false could not be simultaneously true, which is he principle of non-contradiction. To describe the principle of identity we can say that a thing cannot be but itself, although it assumes that if a thing is not itself it violates the principle of non-contradiction where a thing and its negations are simultaneously true. Thus, logic cannot be *described* without using logic. Any attempt to teach logic must use logic, and there is a point at which the learner must see the logical principles intuitively. However, if logic hasn't yet been intuitively understood, then the only method of teaching logic is by using logic.

I will call this the *Principle of Demonstration*: logic can never be *described*, it can only be *demonstrated*. We cannot know logical principles by describing them in terms of things more fundamental. But we see logical principles demonstrated in each description. Thus, we grasp logic not by grasping some more fundamental ideas, but by seeing logical principles in practical use. Ideas underlying logic are elementary and they are grasped by demonstration in descriptions. No amount of thought can intend some object that we might call logic. However, logic can be *demonstrated*, and our knowledge of logic comes not through description but through demonstration.

Once the demonstrability of logic is understood, it is not difficult to see that this applies not just to logic but to every fundamental idea in terms of which we think. The most fundamental ideas cannot be described; they can only be demonstrated. To know something, we must know it in terms of elementary ideas. Knowledge cannot begin with a blank slate as John Locke claimed. To know that there is something out there, we must know how to distinguish it from other things. When we can distinguish things, we can recognize things distinct from one another. The distinctions in terms of which things are known are themselves not reducible. The irreducible ideas should be treated as elementary distinctions from which descriptions are constructed. These descriptions can include thoughts in the mind, theories about matter, and mathematics. And mind, matter and mathematics can then have a common basis in distinctions.

The relation between mathematics and the world was postulated by P. F. Strawson[28] in his search for a metaphysical basis of logical principles. Strawson likened identity, non-contradiction and mutual-exclusion in logic to the mutual separation between objects. An object is itself, and therefore it follows the identity principle. An object cannot simultaneously be another object, and therefore it follows the principle of non-contradiction. An object, if it exists, must be one thing or another, and therefore it follows the principle of mutual-exclusion. The mutual individuality and distinction amongst objects follow the principles of logic and Strawson claimed that logic is derived from the objectivity of the world. The wide application of logic to the world is therefore because logical principles are never violated in the world of objects. Our thinking uses a logical separation between objects. Strawson's point was that logic is not Platonic. Instead the ability to think logically comes from the clarity that exists in the distinguishability of objects. Strawson however incorrectly assumed that the world is distinguished in only one way. This is false because we can divide a non-fundamental set of objects into more fundamental objects. Each division follows the principles of logic, but the principles of logic *underdetermine* the division. If we only had logic, we could infinitely divide the universe into parts. But in the real world, limits to sub-dividing objects are reached when we have reached concepts and actions in terms of which we can think and act. Distinctions thus obey logic, but logic does not exhaust them. Distinctions are essentially the different possible *modes* in which logic operates. There are as many modes as there are elementary ideas and actions, but if there is a limit to divisibility in nature, it should be possible to define the fundamental modes in nature.

We can further understand logic in terms of choices. The principle of mutual exclusion is that we *must* make a choice; choices cannot be avoided; therefore, one side of the distinction must be chosen.

The principle of non-contradiction is that we cannot choose logically contradictory alternatives at once. And the principle of identity is that the choice itself creates a distinction because we choose something and by implication we don't select the alternatives. Logic creates room for choices—because we *must* select one of the alternatives. Logic also *constraints* choices because we cannot pick mutually contradictory alternatives. And logic is itself choice because prior to choice the world only exists as a possibility, and choice converts it into a reality. Therefore, the principle of identity simply says that by choice of A, we have created the reality of A.

By this understanding we can argue that all distinctions are modes of logic, and logic itself is a representation of choice. Thus, the three worlds, and their components, are produced by choice. It's just that choice itself operates in three modes—as relations, universals, and individuals—producing three components of experience. Logic itself operates in many modes to create further distinctions within the three worlds. In that sense, logic alone is insufficient; we must also study the modalities in which it operates. We can call these modalities different kinds of *distinctions* that produce the three worlds and the subdivisions of types within each world.

Once we possess a distinction we can use it as a tool to divide the world in a certain way. The connection between things in the world and thoughts in the mind is that they are built out of the same distinctions. We must have some distinctions in the mind if we are even to see distinct things in the world. Once we see some distinct things, we can try to cognize the world in terms of those distinctions and we will find that they do not always work. To cognize such objects, we must now postulate new types of distinctions. Note that by observing reality we can obtain new distinctions, although we never see the distinction in a single object. If we saw a distinction in a single object, the distinction would be the *description* of that object. But if we only see the distinction in the differences between objects, then the distinction can only be demonstrated by that difference.

Scientific knowledge proceeds by reducing complex things into simple things. This is achieved by postulating distinctions by which we can distinguish complex things as created by simple things. When things cannot be simplified anymore, we have reached the limit to the application of distinctions. The distinctions we used to divide complexity into simplicity forms an elementary basis of distinctions, which represents the most basic ways in which we can use the principles of identity, non-contradiction and mutual-exclusion in nature. Having acquired the distinctions, they become the *tools* by which we can see the world in a new way than we used to before.

The quest for foundations in any field of science is the search for the most basic distinctions. With distinctions, reduction reaches its limits because intellectually there are limits to how elementary we can think. Distinctions are also intuitively self-evident. We cannot forever reduce ideas to something more fundamental because there comes a point at which our thinking cannot go any further and we have hit the most elementary things that we can *think* of. Everything henceforth must be described *in terms of* the elementary distinctions. These distinctions are encoded in language and to describe the world in language we must employ the distinctions. If some distinction does not exist in language, then it cannot be used to describe reality. Indeed, if we are unaware of the distinction we will not even be able to see distinct things. We will only find that our descriptions do not depict the true nature of reality and predictions about reality based on a limited distinction-set do not always work. Our theories will therefore be incomplete, indeterministic, incomputable, and sometimes inconsistent. Every time such problems are encountered, it is time to introduce a new distinction. The set of known distinctions is complete when the descriptions and predictions of nature are complete, determinate, computable, and consistent. We have now reached the point of Occam's razor where we must have no more distinction than those that are totally necessary for science.

Science is the search for the complete set of distinctions. Since distinctions are used to distinguish objects, science is connected to the world of objects. But it is not necessary to have objects to discover the distinctions; the discovery can be pursued within mathematics itself. The difference between mathematics and physics is only that the former searches for distinctions using the mind while the latter, using senses. These methods are complementary because the distinctions that can exist in the mind can also exist in objects. The way we *think* we can divide the world into knowable parts is also the way the world *can* be divided. The only criterion for such a division is that the procedure be conceivable and finite[29]. This implies that the traditional divide between empiricism and rationalism is false. If the distinctions we can think of are the distinctions that can exist in nature, then the mental procedures of mathematics and the physical experiments in physics will reveal the same distinctions.

Every idea can be treated as a concept and as a distinction. When the idea is treated as a concept, it represents a *thing*. When the same idea is treated as a distinction, it acts as a *tool* to divide the world into things. For instance, we can treat the idea of a 'particle' as a concept. We can also treat it as a distinction using which the world can be divided into particles. With every new idea, the world is initially represented as concepts. As these concepts are mastered, the same ideas become tools by which the world is divided. At that point, we think of the world not *as* ideas but *in terms of* the ideas. The difference between the two is that if we think of the world as ideas, then the world is fixed but if we think of the world in terms of ideas, then the world is interpretable. Even if it turns out that something is not an idea, that instance of thinking is falsified but the idea is still useful for other situations. Language conceives the world as ideas such that if something is not a table there could be other things that are tables. We would not discard the idea because one thing is not a table.

The foundations of our thinking are the distinctions by which we divide the world into individuals, and the basic operations by which compute and transform one set of distinct things into another set. These tools may have the same name as the ideas they create, but there is a difference between the tool and the idea it generates. The relation between an idea and a tool is that between a chisel and the statue that is carved by the chisel. An idea is a product of the distinction, like a statue is a product of the chisel. We sometimes think of a chisel as a thing, but that thing is responsible for the making of other things. The foundation of mathematics is similarly in the elementary tools that create ideas. These tools may appear as ideas, but they are more than ideas. The difference between an idea and a tool is that ideas can be reduced, but the tools can't be reduced.

Learning Distinctions

The process by which we acquire concepts illustrates the difference between concepts and distinctions. A concept represents a pure idea, and an object can be labeled by that concept only if the object satisfies all conditions for that concept. In everyday life, however, we label objects by words even when they do not have all the attributes of that concept. When we use these labels in partially fulfilled conditions we are thinking about distinctions and not concepts because distinctions can be applied imperfectly although concepts cannot. Let's illustrate this through some examples. As a concept, a perfectly round thing is a circle. Similarly, as a concept, a square must have straight and equal sides, at right angles to the adjacent sides. Imperfect rounds don't make a circle and unequal lines cannot constitute squares—conceptually or perceptually. However, an object need not be a perfect square or circle for us to call it as a square or circle if these names are treated as *distinctions* from other objects. Thus, the circle vs. square distinction can be used to distinguish even imperfect rounds and uneven squares, and this is how the words are practically used. Philosophers and psychologists tend to believe that in classifying things we think in terms of concepts. Yet the ability to apply labels to objects that don't have all the attributes means that these don't represent concepts but our methods to distinguish.

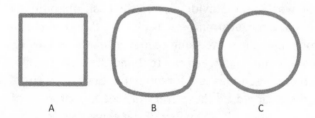

Figure-10 Circle and Square as Distinctions

Consider shapes in Figure-10. If we compare A and B, relatively speaking A is a square while B is a circle. However, if we compare B with C, then relatively speaking C is a circle and B is a square. Whether we label B as a square or a circle depends on whether we are comparing it with a perfect square or a perfect circle. But, in both cases, the same distinction is used to distinguish objects. This ability to apply words even to objects that don't necessarily comply with the perfect criteria means that perfect concepts or percepts do not create knowledge. Knowledge is an outcome of distinctions and not of percepts or concepts. Distinctions are used to create percepts and concepts, and even imperfect circles and squares can be labeled by perfect names. When physical objects aren't present to compare things, we may compare objects in the external world against our memory of past mental representations. A person who knows the meaning of a circle knows how to use the idea of a circle to distinguish objects. If the distinction has been acquired, it can be applied to describing the shape of a table or a plate or a car's wheel.

We teach children the meaning of *square* or *circle* by showing them lots of different squares and circles. Initially, a child thinks that the word 'square' or 'circle' *names* a specific object and a child will argue a circle is something that he or she saw earlier, and the same name cannot be applied to other things. A parent will generally explain that squares and circles are not names, but types of things. Children then begin to think of these as concepts. Now, only perfectly round things are circles and objects with four even and equal sides are squares. Then, the surprise comes: sometimes even things that are not perfectly round or perfectly square can be called by those words. Over time, children learn that these words don't always represent concepts but also distinctions *in terms of* which we must think about objects. To think, a person must acquire distinctions. A child must acquire the distinction of circle versus square. When this distinction is acquired, many worldly objects are circles and squares, not necessarily the one from which the knowledge of the square versus circle distinction was acquired. The distinction has now been internalized and can be applied to the study of the world.

This point can get a little confusing because we use a word like 'circle' or 'square' to denote a percept, a concept, and a distinction. Indeed, most people believe that when we teach children about a circle or a square we are teaching them concepts. However, the concepts of square and circle can be applied only to perfect squares or circles, but distinctions can be used even on imperfect objects.

When children have learnt a distinction, they will talk about their imperfect drawings in terms of perfect ideas. They might call a clumsy shape a square or a circle, and they are speaking about how a shape appears to be relatively squarish or circular in comparison to others. The point of comparison could be the mental image of a square or circle seen earlier. When we call an imperfect round or an uneven square as a circle or a square we are using these not as concepts or percepts but as *distinctions*. We are using a distinction to contrast two things or divide the world into distinct objects *in terms of* the distinction. Distinctions are *modes* in which we distinguish things. Once the transition from name to concept to distinction has taken place in a child's education, the child has acquired a tool by which he or she can divide the world again and again.

Everyday language has several distinctions which may be applied outside the contexts from which they were acquired. Thus, we might contrast a 'healthy' organization with a 'sick' one, a 'well-rounded' person with an 'obtuse' one, a 'bruised' ego with a 'well fed' one, or a 'bitter' experience with a 'sweet' one. We say that some person is 'hot' or 'cool' and 'rough' or 'smooth'. Sometimes a person is 'colorful' while another is 'bland'. In these cases, we haven't touched or tasted a person and found out if their bodies are healthy. But we can distinguish things using distinctions. Similarly, when we call a philosophy or literature or work of art 'tasteless', 'indigestible' or 'wooden', we are not touching, eating or tasting an idea to ascribe those attributes. We are applying a method of contrasting things that is borrowed from one context and applied into another.

Lakoff and Johnson in their classic work *Metaphors We Live By*[30] describe several metaphors latent in everyday language. In these metaphors, distinctions of one domain are applied to distinctions of others, which helps describe experiences better than the vocabulary meant specifically for that subject would. This means that the vocabulary of that domain needs distinctions drawn from other domains. Human creativity is the ability to pick vocabulary from one domain and apply it to another. In such application, we are applying distinctions. The *form* of knowledge remains identical even when distinctions from one domain are applied to another. The distinctions have a name native to a specific domain in which they were originally discovered.

Experiences in some domain may not have yet been refined to an extent to deserve a unique name for it, and so terms from other domains can be borrowed. Language liberates its distinctions from the constraints of the domain that discovers them. It is thus possible to speak of fruity humor and vegetative life.

Distinctions and Mathematics

The fact that we can apply distinctions to label imperfect objects as circles or squares means that a perfect circle or square isn't necessary to do mathematics. With an accurate definition of circle or square, we are expanding the concepts of circle or square in terms of other ideas. Naturally, these descriptions mean that the notion of circle or square are not fundamental to mathematics; at least not as fundamental as those ideas in terms of which things are expanded. The most fundamental things on the other hand cannot be described. They can only be demonstrated. If we must ask what is fundamental, then the answer would be only those things that can be demonstrated but not described. The quest for foundations must therefore end up in the most basic methods or modes used in distinguishing.

But demonstration does not apply only to fundamental things. Things that can be described can also be demonstrated. For instance, the ideas of perfect circles and squares are demonstrated in perfect circles and squares although the distinction of circle and square is demonstrated even in imperfect circles and squares. Some distinctions (that are constructed out of other distinctions) can also be described in terms of the distinctions that constitute them. When mathematics is based on distinctions, it can go beyond current descriptive approaches because mathematics can be applied even to situations that do not descriptively fit the mathematical ideas.

Note how the ability to use concepts 'loosely' gives language its power where we can take an idea from one domain and drop it into another. The key thing in this cross-pollination of domains is that the abstract *form* of the distinction is unchanged from one domain to another. Mathematics deals with abstract forms, although at present these forms are tied to descriptions and something is X only if fits all the attributes of X. When mathematics is treated as the study of abstract forms that are demonstrated, then it can be applied even to situations that only have shades of that distinction but do not instantiate the perfect idea given by that distinction. It would then be possible to use mathematics metaphorically. The fact that a distinction can be used metaphorically does not mean that the distinction does not present to us a pure form. It means that we can see pure forms in things that are not pure instances of that form.

The reconciliation between the pure form and its imperfect instance lies in that a distinction is logical and is comprised of three basic ideas of *identity*, *non-contradiction* and *mutual-exclusion*. Contextual uses exploit the properties of non-contradiction and mutual-exclusion, but not the property of identity. When we distinguish something as a circle by way of a distinction, we may only see *traces* of the perfect roundness in the object. The thing may not be perfectly round, although the distinction represents the notion of a perfectly round. The notion of perfect ideas was first proposed by Plato where we apply perfect ideas from the perfect world to the imperfect objects in the present world by analogy. However, when the mind that holds perfect ideas is part of the imperfect world, then we must first hypothesize a process by which the perfect Platonic world is reflected in the imperfect minds before the mind can see the imperfect world in terms of perfect ideas. This process isn't known today, and the problem is a variation of the mind-body problem where ideas in the Platonic world must interact with the mind before they are represented in the mind. Distinctions do not have this problem. A distinction carries a perfect idea and yet it can be applied even to imperfect situations. The *identity* represents perfection and *mutual-exclusion* and *non-contradiction* the imperfect use.

A distinction implies mutual exclusion, as it separates things. It is the basis for non-contradiction because an assertion and its negation are opposite sides of a distinction. The distinction creates the sense that things are individuated. Since logic does not exhaust the many ways in which things can be distinct, mathematics does not reduce to logic. There are many ways to distinguish objects and each such way represents one kind of identity, non-contradiction and mutual exclusion. Every object in the world abides by the principle of logic, but there are as many ways to apply logic as there are fundamental distinctions by which we could distinguish objects.

Inherent in Strawson's metaphysical view about mathematics is the belief that two things are always different in the same way. The difference between objects is treated as a metaphysical separation because of which they are distinct things and one thing could never be another. But this metaphysical separation leaves no room for interpretations. The world always must be viewed in one way, which is incorrect. This also means that mathematics cannot be used for metaphorical descriptions, unlike ordinary languages.

A strictly descriptive approach leaves out of mathematics the way our minds use and apply abstract forms as distinctions. Strawson's view that logic is the formal structure of all things is right, but his premise that the principles of logic are found in the objective distinct objects is wrong. The converse is in fact true, namely that the objective distinctness of things is based on the application of logical distinctions. An object's uniqueness depends on its distinction from other objects, and we know that distinctness in relation to the other objects. A universe of a single object cannot be known, because to know there must be at least two objects defined by a distinction. Once this principle is generalized and distinctions are the basic features of knowledge, it is obvious that all objects are not distinct in the same way. Rather, the principles of identity, non-contradiction and excluded middle are distinctions, which can be of many kinds.

Distinctions give us the ability to distinguish and once things are distinct then identity, non-contradiction and excluded middle apply. Distinctions can be added, removed, lifted from one domain into another and transformed. We might distinguish the world in ways not done so far or obliterate distinctions that were in use earlier. We might also see shades of a distinction in a situation that hasn't yet used that distinction. Two things that are now distinct may not have been distinct in the past, because their distinguishability is based upon a new distinction that we might have just drawn. Similarly, things that are now distinct may not be so in the future, if we collapse the distinctions that make them distinguishable. All these possibilities are logically feasible and allowed. Mathematics is an extension of the distinction-making and distinction-applying ability.

To know the world, we must first know the elementary ways in which to separate things in the world. These elementary ways are tools by which we divide the world into individual things before we know them. The word 'square' does not just represent the name of the block with which a child has played, but a general idea that can be applied to beds, tables, courtyards, playgrounds and imaginary forms in the sky or uneven drawings on a paper, which may not be perfectly square. A distinction is a *tool* that carves individual forms within a formless world. By looking at distinct things we grasp distinctions, which can then be used for further distinguishing. The foundation of our knowledge rests in the ways in which we can make distinctions. Distinctions can be seen in nature as the basic arrangement that underlies all concepts and percepts. Once we have grasped a distinction we know how to use it to divide the world. Each distinction is a form of logic, and mathematics reduces to logic, although logic has as many forms as there are ways to distinguish.

Methods of distinguishing are ways in which the principles of identity, non-contradiction and excluded middle can be realized. We can represent a distinction as a binary or dyadic relation built from two opposites: for example, hot-cold, rough-smooth, bright-dull, heavy-light, flat-round are dyadic distinctions. When two objects are distinct by the hot-cold distinction, one is hot and the other is cold. Non-contradiction applies contextually to this pair of objects in the sense that if objects are distinct by the hot-cold distinction then they must either be hot or cold. Similarly, excluded middle holds if the distinction is used to distinguish the objects. If two objects are distinct not by hot-cold but by heavy-light, then neither hot nor cold are applicable to the objects, although a new distinction must now be applied. The principles of logic are *context-dependent*, but they are also *context-insensitive*. Every context will use logic in the way it distinguishes objects, although that way differs in contexts.

Fuzzy Logic

With distinctions, we can use a mathematical concept for an object even when it does not satisfy all the object's attributes. Thus, statements that don't fulfill all of concept's criteria can be true, and this is disastrous for mathematics in the traditional sense. Mathematics traditionally presumes that a thing is either X or not, but nothing in between. This comes from the belief that the world is *a priori* distinct things and the separation between X and Y is independent of the mind. Thus, classical propositional or syllogistic logic presumes that a statement like 'this plate is hot' or 'this book is huge' must either be true or false. It implies that something must be either cold or hot, either huge or tiny, etc. But what if the plate is warm and not hot or cold? Or the book is just large and not huge or tiny? What would it do to our claims about plates and books? Will we consider them true or false? Are these statements undecidable?

To solve this problem, logicians invented fuzzy logic, which proposes *degrees* of truth such that we might say that a plate is 'fairly warm' or 'relatively cold' without having to insist on a universal notion about hot and cold. The universal notion of concepts creates a problem in defining their meaning. How much heat will make us call something hot? Whatever definition of hot we have will not make us call something else hot if that object does not have the same amount of heat. Science solves this problem by saying that there isn't anything hot and cold. Instead there are infinite degrees of heat from absolute zero to infinite temperature. But this is problematic because it reduces a concept to its percept and this creates room for logical contradictions as given by Gödel's or Turing's theorems. In fuzzy logic, we create an infinite number of truth values to context-sensitively claim if something is relatively hot or not. So, something can be 0.85 hot and this also defines what is absolutely hot. This is not how everyday contextual usage works. In everyday use we don't say something is 0.85 hot, but just that it is hot. We are not holding notions about an absolute hot in the back of our minds. Everyday contextual usage of distinctions does not contain absolute notions about hot or cold, whereas fuzzy logic depends on absolute ideas.

A state 'fairly warm' or 'relatively cold' can be defined in relation to absolute states 'cold' and 'hot', provided we define the meanings of hot and cold. But, in everyday usage, even if we had two objects in states 'fairly warm' and 'relatively cold' we could still call them 'hot' and 'cold', in a contextual relation. Hot and cold in this case would simply

imply the direction of heat flow and an object can be called hot even though there is no absolute definition of heat. Universal ideas of 'hot' and 'cold' create a problem that necessitates fuzzy logic. However, if concepts and percepts are defined via distinctions, then universal notions of 'hot' and 'cold' are unnecessary. Rather, depending on the context, a statement such as 'the plate is hot' will be true, although in another context the statement would be false.

The causality inherent in the distinction does not depend on a universal standard of a concept. For instance, heat flow from two objects A to B depends on a thermal gradient. Today, this gradient is computed in relation to an absolute scale, but this is not necessary if mathematics would be based on distinctions. The causal behavior about heat flow depends upon a distinction between two objects and not on the absolute scale. It can be studied that way directly.

We need fuzzy logic to address contextual usage only if we assume universal definitions. Distinctions make truth context-sensitive, without violating truth conditions. Universal standards for truth (like a universal definition of 'hot' and 'cold' against a scale) are also part of the distinctions scheme although truth in the case of distinctions will be compared against a universal distinction given demonstratively rather than against a universal standard defined descriptively. We can treat the unit of distinction as a standard scale against which to measure. Both universal and contextual truth is thus accommodated in the distinctions scheme, whereas the universal scheme does not accommodate contextual truth. Fuzzy logic is an anomaly of thinking based upon the notion of absolute and universal truths which lose meaning in contextual uses. Fuzzy logic and its associated problems can be avoided by using distinctions directly.

Semantic Foundations

The idea that the world is built up of distinctions is a shift from universal to contextual knowledge. The shift is not that universals don't exist; it is rather that they exist pairwise (at the least). Thus, the idea 'hot' cannot exist without the idea 'cold'. Therefore, 'hot' and 'cold' are not just universals; they are joined in distinction.

This contextual view is not the same as relativism. This is because contextual knowledge involves the use of universal methods of distinguishing. The universal is now the set of distinctions by which an undifferentiated world can be known. Each person can differentiate things differently, and one could call something hot while another calls it cold. While this difference can be empirically confirmed or denied because the distinctions create objective reality, there is a prior necessity for the difference to be *understood* before the fact can be denied. Underlying the relativism of descriptions is the universalism of a common distinction between hot and cold, which must be understood before the facts are confirmed. While disagreeing on their descriptions, these individuals would still agree on the *language* to be used to express their mutual disagreement.

The context dependence of knowledge means symbols are defined only in relation to other symbols. A *table* is defined by a distinction from chairs, vases, wardrobes, lamps, books, etc. The set of distinctions employed in one context could vary from another context but for people in these contexts to communicate, different sets of distinctions must eventually be mutations of a universal set that everyone can potentially understand. If we think of science as contexts that describe nature, and mathematics as the language used in this description, then mathematics is the collection of distinctions and rules by which distinctions can be transformed. Every area of science can use a mathematical theory of distinctions that gives us the rules by which distinctions are transformed. Different areas of science would describe nature using such a theory. We could see some of the distinctions demonstrated within a theory and in nature, but no theory will exhaust the nature of the distinction itself.

In this sense, mathematics would be independent of its applications, as mathematicians have desired. The mathematical theory of distinctions will also explain why mathematics represents such a good language to describe nature, because it essentially contains the tools that are used to divide and distinguish.

One approach to formulate a complete theory of distinctions is to look at the distinctions across cultures, languages, and domains. These constitute the basic tools by which the world is divided into knowable and usable entities. Ordinary language can be one starting point for collecting distinctions because it embodies distinctions that can be abstracted from our powers of sensation and conception. In one sense, language represents the limits of our minds and thinking, and everything that has been conceived is represented in language. Language contains distinctions that appear repeatedly, and their presence makes language communicable and understandable by others, because everyone is familiar with these distinctions. The key exercise is to identify the set of all distinctions that exist within ordinary language. How distinctions create numbers and higher forms of mathematics trivially results from distinctions because numbers are the products of the process of distinguishing.

Space-Time Distinctions

However, the search for distinctions in ordinary language is a very complex task, given the variety of languages. This task can be simplified by the recognition that no matter what distinctions we come up with, we will have to apply them to divide objects and actions. The division will represent cognitive and computational abilities in matter. Since all objects exist in space-time, the distinctions obtained from language must be applied to space-time to create objects. Locations and directions in space-time are the basis on which we distinguish and count objects. If this counting is based on conceptual distinctions, then the distinctions can ultimately be represented as space-time properties. Meanings can be denoted by space-time properties such as forms, locations, durations and directions. This does not reduce meanings to space-time, although we can represent semantic distinctions using space-time properties.

Note how we have replaced objects and logic in the current foundations with space-time. The notion of a set is similarly modified to denote a *form* in space-time that represents meaning. Distances, durations, locations and directions in space-time represent semantic distinctions. When a set is divided by these distinctions then it creates objects. Objects are by-products of sets or space-time and distinctions. Since sets can be members of larger sets, even objects are forms. Information is the distinctness of objects, but objects come into being by that information. Meaning must exist prior to the objects, and meaning is logically prior to objects. Since meaning can exist both outside and inside matter, the study of meaning does not depend on matter and yet the theory of meaning can serve as a description of matter as well. If mathematics is about meanings, then it is independent of matter although it can be applied to matter.

The key advance needed to formulate this new foundation in mathematics is the ability to treat space-time semantically. In a semantic space-time, locations, extensions, durations and directions represent meaning. Objects in this space-time are symbols of meaning. The categorical mistakes found in Gödel's and Turing's theorems are prevented because space-time properties themselves represent different categories. For instance, the location can denote its conceptual name and meaning. The fact that an object's existence involves no contradictions in space-time can be converted to the idea that the existence of meaning is free of contradictions as well. Space-time does not *a priori* have objects. Rather, objects are created by dividing space-time by an information structure. Since all changes involve finite transformations of meaning in space-time, no mathematical construction can speak about infinite or infinitesimal quantities. Mathematics is Intuitionistic because it involves only finite constructions. Mathematics is Platonic because logical distinctions can exist outside matter. Mathematics is Formalist because when the distinctions exist in matter, mathematics involves only symbol manipulation. Since the same distinctions can exist in the mind and in matter, mathematics is Realist. The semantic approach to mathematics solves the conflict between diverse philosophical positions and logical paradoxes at the same time.

The Philosophy of Tools

Language is a tool that helps us think and do things. As a tool for knowledge, linguistic distinctions help us divide the world into individual objects and use them. Think of a sculptor who carves a statue out of rocks using sculpting tools. He needs a saw for cutting large rocks. For chipping out smaller pieces, he needs chisels and hammers of many varieties. For smoothening surfaces, he needs a grinder. Before a statue comes into existence, the tools must exist to carve it out. In that sense, the tools are logically prior. At present we don't distinguish between tools and objects. We think that tools themselves are products of other tools that carved them prior. And that tools that created the statue-creating tools would have been created using some other tools before. In this way, we think of tools as objects. Science is the discovery of objects and technology makes them into tools, but there is no basic sense in which tools and objects are different. But if all things are the products of tools existing prior, then how did the original tools come about? We must say that there are some fundamental tools using which every other object is constructed. These fundamental tools themselves cannot be created from other tools and they are axioms for any kind of construction.

Gödel treated the alphabets of language as the fundamental axioms of language. The problem is that if we treat alphabets as the fundamental axioms of language, then the categorical distinction between things, concepts, names, instructions, algorithms and problems doesn't exist. It leads to logical contradictions. Our tools must include not just the alphabets but also the modes in which we can combine them to form words and propositions. This approach has other advantages. For instance, once we acquire some tools of thinking, we can export them to other domains, and think of that domain in terms of the tools of distinction discovered previously. Objects can be constructed by carving matter if we know how to apply distinctions. The essential tool for carving things is language. By language I do not mean dictionary words, which are themselves products of distinctions but the distinctions that create them.

Language consists of many distinctions. If we take away some of these distinctions, we take away the ability to divide the world in some ways and thereby of the experiences that arise from those things. If we keep the distinctions but pretend that they don't exist, this leads to paradoxes. In the former case, our world is incomplete because everything that is knowable cannot be known if the distinctions by which we know have been eliminated. In the latter case, our world is inconsistent because we use the distinction but pretend it does not exist. To avoid incompleteness and inconsistency we need all distinctions that make our world complete and consistent. These distinctions are what is *possible* in language.

The linguistic possibilities are preconditions for things that can be created, because we can create only if we can divide things. As a human phenomenon language comes after reality. But language isn't a human phenomenon. If there is a tight coupling between what can exist and what we can think—because minds and matter are both constructed from distinctions—then what is conceivable in language is prior to the human cognition. Distinctions are reflections of what is possible in language and the mind uses these to know reality. We use the same tools to construct objects in the mind as the objects outside the mind. If something possible in language isn't real, then those can be created by using the distinctions. The communication of new things from the mind into matter is through language.

This leads to a basic difference between tools and things, namely that before anything exists, there must be a tool to carve it. Using elementary tools in nature we carve the world into individual objects. Knowledge evolves by creating and obliterating distinctions. New distinctions can be created by combining existing distinctions. Similarly, we can view disparate things as manifestations of the same distinction. Intellectual evolution over the centuries has searched for the most basic distinctions. One modern mistake in this search is the belief that ultimately there is one idea, one theory and one formula using which everything will be described. It is believed that the ongoing unification of scientific theories will collapse the diversity in theories and ultimately there will be only one idea standing. This view is wrong because if we collapse the distinctions, then we will also collapse discussion, debate, and discourse. If we haven't got different positions to take, then we haven't got anything to discuss. Unification will entail silence, because to speak we must distinguish. By throwing away the distinctions, we will throw away the diversity and the myriad experiences that the diversity creates in turn.

Science needs to unify languages and not the content in language. Current science is attempting to unify the content to identify the one idea, theory and formula that governs everything. Instead science should identify the most basic distinctions using which any content can be created. When there is only one idea, then there is nothing left to say. If instead there is unity in language, there are many things to say, which will be understandable by others because they are expressed in a common language. Others may not agree with certain claims, but that disagreement pertains to truth and not to meaning. We might not find some distinction compatible with other distinctions. But that is precisely the nature of distinctions—they have many opposing sides. We might take one of the sides, and by taking sides we are distinguishing ourselves from someone else who takes the opposite position. The unification of language does not take away the ability to choose between alternatives. The foundation of knowledge is in the most elementary distinctions possible in language. Once these possibilities are known, the rest is freedom to use them. Freedom depends on the existence of possibilities and if we collapse alternatives, freedom goes away. If a distinction did not exist, it would restrict the choices that can arise from its use.

By this language we contrast ourselves as black and white, tall and short, rich and poor, cold and hot, sweet and sour, democratic and imperialist, etc. We might prefer one side of a distinction over another or consider one side truer than the other, but this choice does not eliminate the other side of the distinction. To prove our point, we may cite instances in which one side of the distinction is truer than another. And this is the human world of debate, discussion, dialogue and discourse. Different sides of the distinction are useful in different contexts. One tool does not work everywhere, and one distinction cannot be applied universally. By collapsing distinctions, we would lose some part of the discourse. Winning a debate just decides the specific case in which one side of the distinction is better applicable. It does not eliminate the distinction itself. The distinction will come back in another context when the other side is more useful.

There is universalism in this foundation because the tools are accessible to everyone. There is also individualism because tools can be used in combinations, creating interpretation. Universalism makes our creations communicable, because the creations are expressed through distinctions which everyone can understand. And yet we are free to accept or deny distinctions or take opposite sides on a distinction. Our world combines the fixed universal and the free individual. We use the freedom to create novelty and the universal to communicate that novelty. If we only had the universal, we could never have novelty. If we only had novelty, we would be stuck in our imagination unable to tell others about our creation. We need the ability both for creation and communication of novelty and the ability to communicate in a universal language does not preclude the many things that can be communicated using that language.

We are immersed in language and we are the products of language. In fact, we are that language. Every individual who takes a position accesses a part of the possible world and foregoes the opposite parts. He would have to be on the opposite side to know the other possibilities. Being on one side of a distinction is a choice. The collection of all distinctions defines all the choices we can make by deciding to be on one side instead of another. Language thus prescribes the limits of our freedom, the ways in which we can choose. To give up a distinction is to give up a part of that freedom. Learning new distinctions creates more freedom. This freedom has consequences, because when we choose something, we must also reject its opposite. In a world of distinctions, we can never choose everything at once, although we can choose everything one after another. The world can, however, be *known* completely at once by knowing all the distinctions that language facilitates. By knowing the distinctions, we would realize that we can never *experience* it completely at once because experience chooses alternatives one after another. There can thus be a succinct and cohesive representation of the world in a scientific theory that describes the world as all possible distinctions that create the world. However, the theory merely defines the boundaries of what is possible and not what is real. The real world depends on choices that pick a side of the distinction. These choices are not preempted by a scientific theory because a theory can deal with the tools but not with their individual use.

Consciousness and Science

Science grew out of the denial of subjectivity. Descartes divided mind from matter, and science chose to study matter instead of mind. With time, the notion that there is a mind separate from the body has become dogmatic. This has muddled many associated issues about free will, including morality based on which we choose. Paradoxes in mathematics however tell us that a change to science is needed to get to a consistent and complete language. A solution to paradoxes is necessary to allow false statements to be falsified and true statements to be proved. This is the path that holds the greatest promise, scientifically. Before we can speak about free will, we must have room for free will to operate in a logically and mathematically governed world. That possibility does not exist in science today, because the laws of nature predetermine everything. Given an initial state, the final state is fixed by the laws. How can we choose something if the laws of nature determine everything in nature?

The genesis of human free will lies in mathematics in the ability to construct false statements. If the laws of nature permit false statements, then falsity does not result in the inconsistency of the laws themselves. The theories of reality must allow the existence and production of false claims, as well as their denial. Falsities can exist, but that existence does not make them true and there is a difference between a false claim's meaning and its truth. Science can be used to verify whether a belief is false, but prior to that verification, mathematics and logic must allow false beliefs to exist and be produced in a logical system without causing inconsistencies!

Gödel used free will in forming statements whose truth was pending verification. Free will in the context of mathematics is therefore identical to the ability to create meaningful statements that may or may not be true. Free will means that we are free to choose falsities, but we are also free to disprove these falsities using facts or logical consistency. Free will and science have a common meeting point in the improvement of logic to allow for false statements to be computed. This computation represents the *proof* of the falsity but not its *truth*. As we saw earlier, this implies a distinction between proof and truth. Logic is still a tool for thinking, but it is a *tool* in the hands of free will. The computability of false statements does not entail their truth, which means that if the brain can compute a statement it is not necessarily true. Our thoughts must be verified against a reference or use. If mathematics allows false statements to be computed, then there is room for free will. Indeed, the creation of such falsities would constitute an *inappropriate* use of free will.

Epilogue

This book discussed the nature of mathematical paradoxes. These paradoxes arise when category distinctions from ordinary language are imported into mathematics, although mathematics does not have the wherewithal to distinguish between these categories. This inability arises from the fact that mathematics cannot give *meanings* to symbols. Symbols are physical tokens and their meaning is computed through what is called the Arithmetization of Syntax. This technique assumes that symbols are distinct objects but cannot explain how they became distinct. As we saw earlier, to distinguish objects, we must use types. Once objects have been distinguished, they can also be counted which gives rise to numbers. Without supposing the existence of types, mathematics supposes the existence of objects, which is a significant gap not just in mathematics but in our overall view of nature. When we count objects without acknowledging the role that types play in distinguishing objects, we have the freedom to assign arbitrary numbers to them. This arbitrariness is depicted in the choice of coordinate systems to label objects. It is also depicted in the arbitrariness in the *formulae* we use to convert symbols to numbers. Gödel's Numbering is a formula that assigns unique numbers to propositions, but this numbering only pertains to their physical distinctness and not their semantic distinctness. If symbols and propositions are semantically distinct, then the numbering schemes that convert symbols to numbers are indicating only one type of distinction—e.g. the size of the sentence expressing a meaning—not the expression of meaning itself.

There are many ways in which to map symbols to numbers. In some cases, a symbol Not-P can be mapped to the number P. This leads to paradoxes. The paradox could be avoided in two ways. First, if we maintained the categorical distinction between things, names and meanings, the paradox would not arise. But in current mathematics, this distinction cannot be maintained because the categorical distinction requires the idea that objects be distinguished based on types, which implies that types are more fundamental than things, and must therefore have a foundational role in mathematics. The analysis of the categorical confusion therefore only helps us to understand the symptom of the problem, not its fix.

The fix of the problem requires the deeper idea that things are distinct because of types. When we enumerate things, we must enumerate them based on their meanings. Since all things exist in space-time, this requires us to treat space-time semantically rather than physically. In a semantic space-time, the location of an object is its meaning. The object in such a space must be treated as a symbol and its property of location is the 'sound' of that symbol, by which we perceive it. The location thus determines the meaning.

By treating space-time semantically, we not only change how we view objects from things to symbols, but we also prevent the contradictions arising from category mistakes between names, things, and meanings. The succinct moral of this alternative approach is that any theory that treats objects as physical things will be saddled with contradictions. Any theory that treats reality as things rather than symbols is either inconsistent or incomplete. Science can only be complete if it views reality as symbols not things.

Once we arrive at the conclusion that nature is symbols, we must now seek a new foundation of mathematics in which space-time is distinguished into objects by the application of semantic distinctions. These distinctions create a space-time distribution. The distinctness of objects (symbols) is *caused* by semantic distinctions. When we count objects, we must count them based on how they were *constructed* or *created* by applying distinctions to space-time. Such a method of mapping objects to numbers will be free of contradictions that beset current number theory and computing theory.

This idea is further reinforced by problems in geometry which constructs objects and symbols (which are then supposed to denote types) based on typeless points. If the variety of types is indeed a product of a typeless point, then the types are an illusion. But, as we saw above, the notion that types are the products of typeless tokens leads to contradictions. This implies that the type variety cannot be reduced to typeless entities. It must instead be reduced to a fundamental basis of types. The distinction between types creates the distinction between things. When these things are counted based on type distinction, rather than physical properties, we can create a mathematically consistent and complete theory of nature.

Mathematics requires a *Type Number Theory* (TNT) in which objects are distinguished based on types. These objects must be ordered based on types. Counting is the outcome of ordering objects. Types are defined within a collection and therefore set theory has a larger physical role in nature than foreseen so far. Specifically, set theory is the vehicle for creating types by dividing space-time into logically orthogonal parts. Such a theory of numbers will also be a theory of atomic objects and thereby of everything else in nature.

The induction of types into mathematics (and the theory of reality) takes us towards the question about the origin of the types. How are all the basic types created? If nature is not infinitely divisible into points, and any division of matter into objects itself requires fundamental types, then objects are not the *cause* of types; objects are the *effects* of type distinctions. Where are the causes then? This inducts a role for consciousness in mathematics. Consciousness, now, can be viewed as the origin of types. Traditionally, consciousness has been viewed subjectively and inaccessible to objective study. But the question of types frees us from problems of subjectivity. The types that an individual consciousness can potentially understand are common across all individuals, and this forms the basis of our ability to communicate with others. An individual only chooses to use some types as opposed to others. The universal set of types is the ways in which a consciousness can use its choices. These types represent the *language* that consciousness can potentially use and understand. The study of consciousness that has traditionally been mired in problems of privacy and subjectivity can now be objectified as the study of the fundamental meanings that consciousness can *potentially* understand and use. These meanings are present in matter and frequently employed in everyday language. They include (but are not exhausted by) the categories of naming, meaning, things, instructions, algorithms, and problems. They also include distinctions that create various distinct symbols and propositions.

In a sense, the foundation of the world is language. The study of language is in one sense a study of consciousness. In another sense, the theory of language is also the theory of numbers, physical objects and every other type of natural reality. If this language can be mathematically described, then mathematics will be the queen of sciences that explains the production of all things in the universe.

Endnotes

1. MECHANIZING THOUGHT

1 In modern theorem proving machines, intelligence is not identifying or grasping an idea as being different from other ideas, but searching for the simplest possible way in which a pre-defined solution to a problem can be constructed.

2 Mathematical Problems (lecture delivered before the International Congress of Mathematicians at Paris in 1900). Translated by Dr. Maby Winton Newson into English and published in the *Bulletin of the American Mathematical Society* 8 (1902), 437-479.

3 Kripke, Saul (1980), "Naming and Necessity", Cambridge, Massachusetts: Harvard University Press.

4 Before Gödel proved the incompleteness of mathematics, he had showed that logic is both consistent and complete. This is called Gödel's completeness theorem. Together, the completeness and incompleteness theorems imply that logic is consistent and complete, while number theory is either consistent or complete. The source of incompleteness therefore lies in the theory of numbers.

5 Kleene, Stephen (1952), "Introduction to Metamathematics", North-Holland.

6 This applies only to Turing Machines. In *neural algorithms*, an algorithm is incrementally tweaked until it attains desired results. But the tweaking of the algorithms isn't based upon an understanding of what is being changed and the rationale behind why a change is expected to fix the gap.

Algorithm tweaks are random changes to a procedure intended to check if that helps solve a problem. Such procedures are more in the vein of trial and error changes intended to see if some such trial works out. The approach is inspired by evolutionary theories in biology where random changes are supposed to improve an organism's chances of survival. Such changes are not based upon an *understanding* of the problem and what will help us get closer to a solution.

2. GÖDEL'S MISTRICK

7 "The Philosophy of Logical Atomism", reprinted in The Collected Papers of Bertrand Russell, 1914-19, Vol 8., p. 228.

8 The fundamental theorem of arithmetic is that any integer greater than 1 can be written down as a product of prime numbers. Prime numbers are those numbers that are divisible only by 1 and by themselves. Examples of prime numbers are 2, 3, 5, 7, 11, etc. as they can either be divided by 1 or by that number itself.

9 Gödel's numbers are known to not be unique, but the points of non-uniqueness are few, and this overlap has no bearing on the conclusion of his theorem. Specifically, it is not just the non-unique numbers at the root of the proof of incompleteness; even unique numbers also result in the same conclusion.

10 Not every grammatically correct statement is meaningful, but that's a separate issue, namely that syntactical rules aren't enough to ensure a statement is meaningful. Let's sidestep that issue and assume that all grammatically correct statements are meaningful. This assumption is generally false in ordinary language, but it is true in mathematics.

11 'Logic' here refers to first-order logic. Second and other higher-order logics can refer to other statements. First-order logic is complete, but second and higher-order logics suffer from Gödel's incompleteness.

12 There is little consensus today on which amongst machine or human methods of checking program correctness is better. For instance, some

programmers believe that one should do "code reviews" before any testing to quickly catch problems without expending efforts in testing. Other programmers believe that testing must be performed before code reviews so that the reviewer knows that the program more-or-less works for most cases and the review must only catch corner cases.

13 It is estimated that the total number of atoms in the universe is within the range of 10^{78} to 10^{82}.

3. MATHEMATICS AND REALITY

14 This idea is more fully described in my other work entitled *Quantum Meaning: A Semantic Interpretation of Quantum Theory.*

15 The modern theory of space-time from General Relativity suffers from the problem that it is possible to perform some matter redistributions (called Active Diffeomorphisms) such that object trajectories now intersect at a different point in space. This would seem to imply that General Relativity is indeterministic. To solve this problem Einstein used what is now called the Point Coincidence Argument where space-time is *defined* by the intersection of trajectories. We don't therefore speak about the intersection at a different point, because the point is the intersection of trajectories. Einstein believed that there is no meaning to space-time other than through intersection of object trajectories. The philosophical import of this idea is that space-time is not logically prior to matter, and if we removed all matter from space-time, there will be no space-time. In other words, the concept of 'empty' space-time is vacuous in General Relativity.

16 A Dedekind cut is a partition of a totally ordered set into two non-empty parts, (A, B), such that A is closed downwards (meaning that for all a in A, $x \leq a$ implies that x is in A as well) and B is closed upwards, and A contains no greatest element.

17 This point about atomic reality is extensively covered elsewhere in my work on the Semantic Interpretation of Quantum Theory, *Quantum Meaning*. But briefly, a semantic interpretation of atomic theory tells us that particles in an ensemble are not many different things of the same type,

because, logically speaking, a set can only contain one member of a given type. Classical ensembles are not sets because they allow multiple objects of the same type (they are all essentially particles). Within a set, there can be only one member of one type. When this idea is applied to a quantum ensemble (as different from a classical ensemble) we arrive at the idea that members of a quantum ensemble are different types of objects.

4. NUMBERS AND MEANINGS

18 "Zeno's Paradoxes", Stanford Encyclopedia of Philosophy. Retrieved online at https://plato.stanford.edu/entries/paradox-zeno/.

19 Tussy, Alan; Gustafson, R. (2012). "Elementary Algebra" (5th ed.), Cengage Learning.

20 The spherical coordinate system is a prominent example. It uses three coordinates—r, θ, and φ—the last two of which are directions.

21 This statement is true when we think of the interaction between objects, but untrue when we think of *why* two objects interact (as opposed to other objects). In atomic theory, for example, the interaction between objects is given by a probability, which means that some objects interact far more than others (this probability is the basis of the well-known measurement problem in quantum theory). Probabilities can be attributed to a macroscopic object (an ensemble of particles) if we said that they have natural *tendencies* to interact relatively more or less with others. For example, you may have the ability to read and eat, but you may spend more time eating than reading. This is because of an innate *nature* which determines your preference to use one ability far more than others. The individuality therefore has causal effects.

22 This idea is more extensively discussed in my book *Signs of Life* where I describe the genesis of species from a world of ideas, not through gradual evolution from preexisting species. Every species is eternal as a possibility, but it becomes real at different times when universality and individuality are combined. This idea stems from the universality of ideas. A species in this view is an idea that preexisted its manifestation into the individual

members of the species. So, the species is never created or destroyed; the individual members are created or destroyed.

23 Cantor's fundamental insight was that natural numbers and real numbers were not *equinumerous*. That is, the size of the set of natural numbers was much smaller than the size of the set of all real numbers.

24 Gauss, Carl Friedrich; Clarke, Arthur A. (translator into English) (1986). "Disquisitiones Arithemeticae" (Second, corrected edition). Springer..

25 "Goldbach Conjecture". MathWorld website. Retrieved at: https://mathworld.wolfram.com/

26 Infinite series expansions of many irrational numbers are known to exist. These expansions can be coded as finite programs, but each step in the series itself requires the computation of an infinite digit rational number. An example of π is shown below.

$$\pi = \frac{4}{1} - \frac{4}{3} + \frac{4}{5} - \frac{4}{7} + \frac{4}{9} - \frac{4}{11} + \cdots$$

Such a method for computing irrational numbers decomposes a program with infinite programmatic information into two or more programs—each with a finite amount of programmatic information. However, if we were to execute such a program, the execution would be perpetually stuck at the first step that involves the computation of an infinite rational number (4/3 in the example). Of greater importance is the fact that these series only represent *approximations* because the sum of rational numbers is always a rational number and never an irrational number. Thus, while we can write a finite approximate program for an irrational number this program will not truly represent an irrational number.

5. MATHEMATICAL FOUNDATIONS

27 In logic, the terms extension and intension are used as two types of meanings. The term "intension" indicates the internal meaning of a term or concept that constitutes its definition. The term "extension" indicates the

range of applicability of that idea by naming the objects to which it applies. For instance, the intension of "car" is "vehicle for conveyance on streets," whereas its extension embraces such things as sedans, SUVs, sports cars and family cars.

28 Strawson, P. F. (1952), "Introduction to Logical Theory", London: Methuen.

29 Mathematicians have had a history of postulating infinite procedures. Thus, mathematics conceives infinite series, infinite division to construct infinitesimals, and then infinite additions of infinitesimals to construct finite quantities. Such infinite and infinitesimal procedures are not computable, although they are conceivable. An algorithm that prescribes an infinite procedure will not terminate.

30 Lakoff, George and Johnson, Mark (1980), "Metaphors We Live By", The University of Chicago Press.

Index